T0344897

TV White Space

TV White Space
The First Step Towards Better Utilization of Frequency Spectrum

Ser Wah Oh
Yugang Ma
Ming-Hung Tao
Edward Peh

The ComSoc Guides to Communications Technologies
Nim K. Cheung, *Series Editor*
Thomas Banwell, *Associate Series Editor*
Richard Lau, *Associate Series Editor*

Copyright © 2016 by The Institute of Electrical and Electronics Engineers, Inc.

Published by John Wiley & Sons, Inc., Hoboken, New Jersey. All rights reserved
Published simultaneously in Canada

For general information on our other products and services or for technical support, please contact our Customer Care Department within the United States at (800) 762-2974, outside the United States at (317) 572-3993 or fax (317) 572-4002.

Wiley also publishes its books in a variety of electronic formats. Some content that appears in print may not be available in electronic formats. For more information about Wiley products, visit our web site at www.wiley.com.

Library of Congress Cataloging-in-Publication Data is available.

ISBN: 978-1-119-11042-2

Printed in the United States of America

10 9 8 7 6 5 4 3 2 1

Contents

5.   Worldwide Deployment   **165**

Preface

TV white space (TVWS) is gaining a lot of attention recently, not only due to its ability to bridge the gap in certain applications but also due to its innovative ways of using spectrum that could radically change the way how spectrum is being allocated and used in the future. Many people asked me to recommend a good reading on this topic and I have to always send a list of articles that are not connected. Although there are some books available, those books are only providing collection of first-order information that requires the readers to link the dots on their own. Coincidentally, IEEE–Wiley approached me to write a book and I decided that it is the right time to have a book that could help the readers to understand this topic better.

I started working on TVWS since more than a decade ago, at a time when nobody knew whether TVWS works or not. Things started to make a turn when the Federal Communications Commission (FCC) concluded a successful trial in 2008 where I had the honor to be part of the parties participated in that trial. I still remember I had to fly 24 hours from Singapore to Maryland to submit and test our prototype. At that time, we were the only non-USA based organization participating in that cornerstone trial together with Microsoft, Motorola, Philips, and a Silicon Valley start-up Adaptrum.

With regulation in the United States ready, Microsoft approached me to expand the reach to other regions. Singapore was a natural choice as it was leading the TVWS regulatory efforts in the Asian region. We decided to start a group that brought the ecosystem together called the Singapore White Spaces Pilot Group (SWSPG) where I assumed the role as Co-Chairman of the group together with Microsoft and StarHub (Singapore's second largest telecom operator). In the inaugural international workshop organized by SWSPG where I chaired, I shared the TVWS ecosystem using the famous business framework invented by a management guru called Michael Porter's Six Forces.[1] Porter looks at an industry from suppliers, buyers, competitors, new entrants, substitutes, and complementors perspectives that formed the six forces. Each force has different levels of influence to the industry that needs to be studied thoroughly in order to position oneself correctly in the value chain. I attempt to use this same framework to guide the organization of this book for ease of understanding the linkages among the different "forces."

---

[1] Note that the original framework was called Porter's Five Forces, but it has since been expanded to Six Forces.

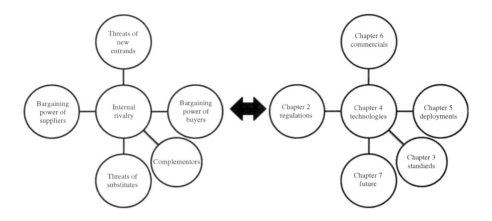

Chapter 1 lays down the background information and related activities with respect to TVWS. From TVWS perspectives, regulators are the "supplier" of spectrum. The power of supplier, that is, regulation that has strong influence on the success of TVWS is introduced in Chapter 2. Chapter 3 looks at the complementors to TVWS ecosystem. Very much like doctor who is a complementor to medicine, standards tend to influence one's choice of technology solutions. Different players compete for market shares in this industry by introducing and advancing technologies and solutions, which are discussed in Chapter 4. Chapter 5 addresses the demands from the buyers' perspectives. Various use cases and deployments based on the needs of the buyers will be discussed in this chapter. Chapter 6 studies the market and commercial potential of TVWS and other spectrum sharing technologies. How attractive this market is will determine whether new entrants are keen to enter this market. Finally, Chapter 7 attempts to predict some future trends related to this technology. The future will determine whether the technology continues to flourish or substitutes will enter and compete.

We hope by painting this book into a pictorial framework will ease the readers in understanding the various aspects of TVWS. Finally, the readers should bear in mind that TVWS is just one of the earliest systems that utilizes dynamic spectrum access mechanism where secondary users coexist with primary users. This mechanism will change the way how spectrum is being used in the future. Watch this space!

Ser Wah Oh

Abbreviations

ACLR	Adjacent Channel Leakage Ratio
ATU	African Telecommunications Union
CAK	Communications Authority of Kenya
CAPEX	Capital Expenditures
CP	Cyclic Prefix
CR	Cognitive Radio
CRN	Cognitive Radio Network
DAB	Digital Audio Broadcast
DMB	Digital Multimedia Broadcast
DTT	Digital Terrestrial Television
DVB	Digital Video Broadcasting
DVB-H	Digital Video Broadcasting—Handheld
DVB-T	Digital Video Broadcasting—Terrestrial
DSA	Dynamic Spectrum Access, Dynamic Spectrum Alliance
GDD	Geolocation Database Dependent
FBMC	Filter Bank Multi-Carrier
FDMA	Frequency Division Multiple Access
FFT	Fast Fourier Transform
FIC	Fast Information Channel
GFDM	Generalized Frequency Division Multiplexing
GLN	Gigabit Libraries Network
GPS	Global Positioning System
HPC	High-Priority Channels
HTTPS	Hypertext Transfer Protocol Over Transport Layer Security
HVAC	Heating, Ventilation, and Air-Conditioning
IFFT	Inverse Fast Fourier Transform
IoT	Internet of Things
IPM	Interference Power Management
ISM	Industrial, Scientific, and Medical
ITB	Interoperability Test Bed
M2M	Machine-to-Machine

MAC	Medium Access Control
MICS	Medical Implant Communication Service
MIMO	Multiple-Input, Multiple-Output
NLOS	Non-Line-of-Sight
NPV	Net Present Value
OFDM	Orthogonal Frequency Division Multiplexing
OFDMA	Orthogonal Frequency Division Multiple Access
OPEX	Operational Expenditures
PAD	Programme-Associated Data
PAWS	Protocol to Access White Space
PMSE	Program Making And Special Events
PPDR	Public Protection And Disaster Relief
PRT	Personal Rapid Transit
PSD	Power Spectral Density
PSI/SI	Program-Specific Information/Service Information
PWMS	Professional Wireless Microphone Systems
QoS	Quality-of-Service
REM	Radio Environment Map
RF	Radio Frequency
RLSS	Registered Location Secure Server
RSM	Radio Spectrum management, New Zealand Telecommunication Authority
RTS/CTS	Request to Send/Clear to Send
SDMA	Space Division Multiple Access
SFN	Single-Frequency Networks
STOD-RP	Spectrum-Tree Based on-Demand Routing Protocol
TDMA	Time Division Multiple Access
TDOA	Time Difference of Arrival
TVWS	TV White Space
UHF	Ultra-high Frequency
URI	Uniform Resource Identifier
VHF	Very High Frequency
WSDB	White Space Databases
WSDB-Q	WSDB with QoS
WSDIS	White Space Devices Information System
WSD	White Space Devices

Chapter 1

Introduction to Cognitive Radio and Television White Space

Are wireless communication systems smart?

Wireless communication is a key enabler for smart systems such as smart cities, autonomous ports, and so on. However, are the current wireless communication systems themselves smart? This book looks at some initial aspects of smartness in wireless communication that could pave the way for future smarter wireless communication systems.

In the past decades, the demands for wireless communications grew exponentially. There is no sign of slowing down; instead, the growth seemed to be accelerating. Besides traditional television and audio broadcasting, today wireless communications are used everywhere from public cellular phone systems to enterprise wireless networks, as well as specialized systems in railway, maritime, and aviation transports, medical electronics, remote sensing, emergency services, security surveillance, military, radio astronomy, satellite, and so on.

Due to the emergence of Internet of Things (IoT), wireless applications are rapidly expanding into machine-to-machine (M2M) connections. All these applications make extensive use of the frequency spectrum. Many of such applications are bandwidth hungry. To cater for the growth of various new services, many countries all over the world have crafted their national broadband plans to address future communication needs with the aim to stimulate economic growth by providing better information infrastructure. Among the broadband alternatives, wireless is one of the most important items in these plans.

Unfortunately, spectrum is a scarce resource. While technologies are improving to squeeze more data into each frequency channel, policy makers still face stiff challenges in addressing the needs of everybody since the demand (in bandwidth) outpaced that of supply. For services that require dedicated channels, for example, cellular, it becomes harder and harder for regulators to clear up a chunk of

TV White Space: The First Step Towards Better Utilization of Frequency Spectrum, First Edition.
Ser Wah Oh, Yugang Ma, Ming-Hung Tao, and Edward Peh.
© 2016 by The Institute of Electrical and Electronics Engineers, Inc. Published 2016 by John Wiley & Sons, Inc.

frequency bands for these services. For example, the regulators worldwide find it hard to fully harmonize the spectrum for the fourth generation (4G) Long Term Evolution (LTE) networks since different countries have different fragments of spectrum left for such assignment. This issue is becoming more and more prominent as can be seen in the proposals toward the fifth generation (5G) networks where there is simply no common frequency spectrum available.

On the other hand, regulators also assigned some frequency spectrum as unlicensed bands such as the industrial, scientific, and medical radio bands (ISM). In these unlicensed bands, different wireless systems share the same spectrum resource and coexist by adopting politeness protocols. Wireless systems that use unlicensed bands, such as Wi-Fi, Bluetooth, ZigBee, and so on, are becoming more and more attractive since they have better spectrum utilization rate and there is no charge in spectrum usage.

One might ask why not the regulators assign all spectrums to be unlicensed? This book attempts to shed some light on the various considerations for spectrum usage.

In this chapter, we look at the various aspects of the spectrum, including their utilizations (Section 1.1), the future needs (Section 1.3), and why do we need more agile and dynamic spectrum access technologies (Sections 1.4 and 1.5).

Taxonomy

In this book, readers will see terms such as spectrum, frequency, band, channel, bandwidth, data rates, and so on appearing quite frequently. In some instances, they may be used interchangeably. In other cases, they refer to slightly different phenomena. In order to help the readers understand the differences, we use another natural resource, that is, land, as analogy.

Spectrum	Land
Frequency	Different types of lands, hilly, flat, forest, wetland, and so on
Band	Division of land for different usage, for example, parks, housing, roads, and so on
Channel	Marking of land so that rules could be followed more easily, for example, lane markings on roads
Bandwidth	Using the road as an example, this will be how wide the lane is being marked, for example, lanes for cars on highway are much wider than lanes for motorcycles
Data rates	If you view different vehicles as data packets, this is how fast the vehicles can flow through the road

1.1 SPECTRUM SURVEY

Let us take a look at the current spectrum utilization. Figure 1.1 shows the frequency allocations from 30 MHz to 300 GHz in the United States. It can be seen that the spectrum in this frequency range has been fully assigned.

On the other hand, if we take a look at the spectrum utilization of the assigned frequencies in Table 1.1, which is the spectrum occupancy measurement at Green Band, West Virginia, we will be surprised that the utilization rate of the assigned frequencies is so low, and the average spectrum use is just 1% from 30 MHz to 2.7 GHz. There is a similar measurement report shown in Figure 1.2, where we can see that the spectrum efficiency is about 6.5% from 30 MHz to 7 GHz in Singapore. This implies that there are plenty of rooms for improvement. Spectrum is a natural and finite resource that all wireless communications rely on. Only if we continuously increase the spectrum utilization can the growth of wireless communications be sustainable.

1.2 SPECTRUM HARMONIZATION

As we know, the spectrum can be used according to different frequencies, different time, and different locations, namely, frequency division, time division, and space division. If all assigned spectrums are used by a single user, the utilization of spectrum can be synchronized well. Perfect frequency, space, and time divisions can be achieved. However, the challenge is that different frequencies are used by different users with different wireless communication systems, including cellular, broadcasting, navigation, transports, medical, remote sensing, satellite, and so on. Moreover, these systems upgrade along technology improvement independently. How to coordinate them becomes an issue.

Due to potential interference among different spectrum users, many portions of the radio spectrum are controlled through regulations by various governments. Although the individual national government has the right to make their own radio regulations, the radio signal and frequency usage do not stop at national boundaries. It would be much more economical and convenient for the users, manufacturers, and suppliers if common frequencies are used for the same services in different countries. Thus, regulations of frequency usage for various applications among the geographically neighboring countries or even across the world are usually intentionally coordinated by these regulators. The pioneering regulations from more advanced countries are usually followed and adopted by the other countries. For example, the regulations from the Federal Communications Commission (FCC) of the United States are widely used by many other countries.

Besides regulations, standardizations are also required for efficiently utilizing the limited resources by establishing a basis for harmonized use of spectrum. This is achieved by setting appropriate technical specifications, such as radiation power, bandwidth, emission limits, air interface, communication protocol, and so on.

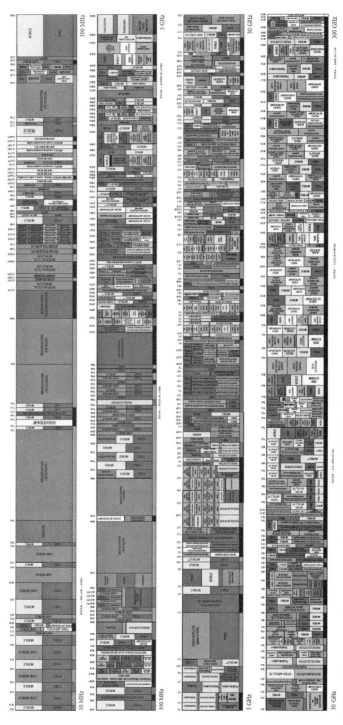

Figure 1.1 A typical frequency allocation from 30 MHz to 300 GHz. (The chart was produced by the U.S. Department of Commerce, National Telecommunications and Information Administration, Office of Spectrum Management, January 2016.) (From Ref. [1]. Public domain.)

Table 1.1 Summary of Spectrum Occupancy in Green Bank, West Virginia [2]

Start Frequency (MHz)	Bandwidth (MHz)	Spectrum Band Allocation	NRAO Spectrum Fraction Used	NRAO Occupied Spectrum (MHz)	Average % Occupied
30	24	PLM, Amateur, others	0.00045	0.01	0.0
54	34	TV 2–6, RC	0.11056	3.76	11.1
108	30	Air Traffic Control, Aero Nav	0.15485	4.65	15.5
138	36	Fixed Mobile, Amateur, others	0.02745	0.99	2.7
174	42	TV 7–13	0.00220	0.09	0.2
216	9	Maritime Mobile, Amateur, others	0.00556	0.05	0.6
225	181	Fixed Mobile, Aero, others	0.01842	3.33	1.8
406	64	Amateur, Radio Geolocation, Fixed, Mobile, Radiolocation	0.00379	0.24	0.4
470	42	TV 14–20	0.00379	0.16	0.4
512	96	TV 21–36	0.04283	4.11	4.3
608	90	TV 37–51	0.00156	0.14	0.2
698	108	TV 52–69	0.00113	0.12	0.1
806	96	Cell phone and SMR	0.00017	0.02	0.0
902	26	Unlicensed	0.00004	0.00	0.0
928	32	Paging, SMS, Fixed, BX Aux, and FMS	0.02459	0.79	2.5
960	280	IFF, TACAN, Global Positioning System (GPS), others	0.00000	0.00	0.0
1240	60	Amateur	0.00012	0.01	0.0
1300	100	Aero Radar, military	0.00000	0.00	0.0
1400	125	Space/Satellite, Fixed Mobile, Telemetry	0.00000	0.00	0.0
1525	185	Mobile Satellite, GPS L1, Mobile Satellite, Meteorological	0.00082	0.15	0.1
1710	140	Fixed, Fixed Mobile	0.00000	0.00	0.0
1850	140	PCS, Asyn, Iso	0.00001	0.00	0.0
1990	120	TV Aux	0.00009	0.01	0.0
2110	90	Common Carriers, Private Companies, MDS	0.00353	0.32	0.4
2200	100	Space Operation, Fixed	0.00000	0.00	0.0
2300	60	Amateur, WCS, DARS	0.10521	6.31	10.5
2360	30	Telemetry	0.00004	0.00	0.0
2390	110	U-PCS, ISM (unlicensed)	0.00007	0.01	0.0
2500	186	ITFS, MMDS	0.00137	0.26	0.1
2686	214	Surveillance radar	0.00288	0.62	0.3
Total	2850			26.14	
Total available spectrum				2570	
Average spectrum use (%)				1.0%	

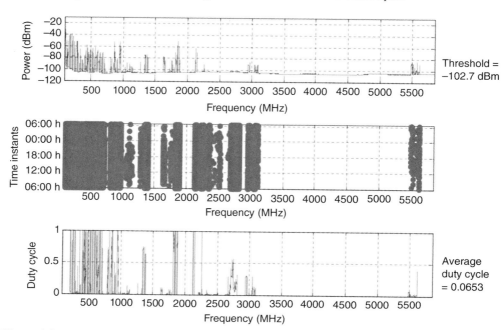

Figure 1.2   Spectrum utilization from 30 MHz to 6 GHz in Singapore. (Reproduced from Ref. [3] with permission from IEEE.)

On the spectrum coordination and standardization for wireless communications, the international standards bodies play important roles. At the global level, the International Telecommunication Union (ITU) is the flagship organization, which seeks global coordination of spectrum usage. Its World Radiocommunication Conference (WRC) is taking place approximately every 4 years or whenever necessary to review and revise the multinational global radio regulations governing the use of the radio frequency spectrum and also the geostationary/nongeostationary satellite orbits. ITU also holds Regional Radiocommunication Conference (RRC), which covers region and group of countries, to develop agreements concerning particular wireless communication services or frequency bands. The decision in RRC could not modify the global radio regulations, but has binding effect on those countries that are party to the agreement. The Institute for Electrical and Electronics Engineers (IEEE), as the world's largest professional association, is also active in globally standardizing wireless communication technologies. One successful example of global standards from IEEE is the 802.11 series of standards for Wireless Local Area Networks (WLAN), which are widely adopted in the world.

At the regional level, European Telecommunications Standards Institute (ETSI) is a well-known organization making telecommunications standardizations for European Union. Its European Conference of Postal and Telecommunication Administrations (CEPT) aims to coordinate spectrum use in the European Union.

In the European Union, the regulatory framework is defined in the Radio Spectrum Policy Decision (Decision No. 676/2002/EC), which seeks to

- link spectrum demands to EU policy initiatives,
- bring out legal certainty for technically unified measures carried out by CEPT,
- enhance transparency and unified format of information on the use of spectrum by member states, and
- promote European interests in international negotiations.

In the European Union, Radio and Telecommunications Terminal Equipment (R&TTE) follows R&TTE Directive, which obeys harmonized standards developed by ETSI according to the request of the European Commission. These standards cover technical characteristics, including effective use of the radio spectrum as well as interference avoidance. Equipment manufactured in accordance to ETSI Standards is allowed to get into the EU market. Network operators cannot decline connecting compliant equipment with ETSI standards. In the European Union, spectrum management remains a national matter, and authorities of the member states are allowed to regulate radio interfaces, in the terms of the Directive, and the member states need to publish their regulations in the same format.

Japan is another country where they publish and adopt their own standards for many wireless communication systems. This situation changed when Japan harmonized their third generation (3G) cellular standard with that of ETSI in the late 1990s. In contrast, as China is growing to become the world's second largest economy together with the world's largest population, it has started to define its own standards. Most other countries in the world tend to follow the regulations and standards defined by the larger economies.

An obvious trend of the spectrum management is the growth of license–exempt spectrum. In the past, exclusive licenses for specific frequency bands and specific purposes dominated government spectrum policy. In the last decade, however, governments around the world have embraced the concept of "license–exempt" (also known as "unlicensed") spectrum as another way to encourage their citizens to adopt innovative new wireless technologies. License–exempt spectrum refers to frequency bands, such as those bands used for Wi-Fi technologies, for which regulators do not grant exclusive licenses, but instead protect against interference and achieve important operational safeguards through equipment certification and clear and enforceable technical rules.

The result of this shift is dramatic. License–exempt spectrum has been, and continues to be, a powerful catalyst for innovation and investment. Today, there are over 10 billion devices connected to the Internet, and connections leveraging license–exempt spectrum access carry the majority of Internet traffic.

1.3   NATIONAL BROADBAND PLAN

To meet the exponentially growing demands on high-speed communications, many countries crafted their own national broadband plan (NBP). For most countries, wireless access is a key focus.

1.3.1 The United States

In the United States, the FCC released its NBP on March 17, 2010 by setting out the roadmap for initiatives aiming to stimulate economic growth, increase jobs, and boost U.S. capabilities in health care, education, homeland security, and even more [4,5]. The plan includes the following:

1. Promote world-leading mobile broadband infrastructure and innovation, which is a plan to make an additional 500 MHz of spectrum available for mobile broadband within the next 10 years through unleashing more spectrum for mobile broadband usage; increasing opportunities for innovative spectrum access models; removing barriers to spectrum utilization; improving data and transparency regarding spectrum allocation and utilization.

2. Accelerate universal broadband access and adoption, and advance national purposes such as education and health care. This is a plan to provide an array of recommendations to accelerate universal broadband access and adoption—including for rural use; low-income Americans; schools and libraries; hospitals, clinics, doctors and patients; Americans with disabilities; and native Americans—and to advance national purposes such as education, health care, and energy efficiency.

3. Foster competition and maximize consumer benefits across the broadband ecosystem, which is a plan containing several recommendations to promote competition and empower consumers across the markets that make up the broadband ecosystem: network services, devices, and applications, through removal of barriers to entry by streamlining access to key broadband inputs; improve data collection, analysis, and disclosure to promote broadband competition and protect and empower consumers; and unleash innovation and competition in video devices.

4. Advance robust and secure public safety communications networks, which is a plan recommending a series of actions to help ensure that broadband can support public safety and homeland security, respond swiftly when emergencies occur, and provide the public with better ways of calling for help and receiving emergency information, through facilitating creation of a nationwide interoperable public safety mobile broadband network; promoting cyber security and protect critical communications infrastructure; and promoting development and implementation of next-generation 911 and alerting systems.

1.3.2 Canada

In Canada, NBP started in March 2009. Canada was one of the first countries to implement a connectivity agenda geared toward facilitating Internet access to all of its citizens. However, gaps in access to broadband remain particularly in rural and remote communities. In Canada's NBP, it is planned to fill up the broadband gap by encouraging the private development of rural broadband infrastructure.

1.3.3 The European Union

In the European Union, the NBP is in the form of Digital Agenda, which is one of the seven flagship initiatives of the Europe 2020 strategy. The objective is to bring "basic broadband" to all Europeans by 2013 and also to ensure that, by 2020, all Europeans have access to much higher Internet speeds of above 30 Mbps with 50% or more of European households subscribing to Internet connections above 100 Mbps. In September 2010, the European Commission published a "Broadband Communication," which describes the measures the Commission will take to achieve the targets of the Digital Agenda.

1.3.4 The United Kingdom

In August 2009, the UK Government published its Digital Britain Implementation Plan setting out the government's roadmap for the rollout of its plans mentioned above [6].

Next-generation broadband networks will offer not just conventional service but also more revolutionary applications, including tele-presence, allowing for much more flexible working patterns, e-health care in the home, and for small businesses the increasing benefits of access to cloud computing, which substantially cuts costs and allows much more rapid product and service innovation.

Next-generation broadband will enable innovation and economic benefits that we cannot predict today. First generation broadband provided a boost to GDP of some 0.5–1.0% a year. In recent months, the United Kingdom has seen an energetic, market-led roll-out of next-generation fixed broadband, which will inevitably increase the demands for wireless.

Following decisions by the regulator, Office of Communications (Ofcom), which have enhanced regulatory certainty, British Telecommunications (BT) Group has been encouraged by the first year capital allowances measures in Budget 2009 and the need to respond competitively to accelerate their plans for the mix of fiber to the cabinet and fiber to the home. BT's enhanced network will cover the first 1,000,000 homes in their network. The £100m Yorkshire Digital Region programme approved in Budget 2009 will also provide a useful regional test bed for next-generation digital networks.

1.3.5 Japan

The Japanese Cabinet released the e-Japan Priority Policy Programme in 2001. In this programme, the private sector plays the leading role in information technology, and the government is to implement an environment in which markets function smoothly through the promotion of fair competition and removal of unnecessary regulations. The e-Japan program extended tax incentives and budgetary support for carriers building advanced communication infrastructure.

Under this programme, there are two policies: the National Broadband Initiative, which mandates that federal and local governments deploy fiber to underserved areas; and the e-Japan strategy, which set forth the goal of providing access at affordable rates by 2005 to high-speed Internet networks for at least 30 million households and to ultrahigh-speed Internet networks for 10 million households.

These policies provided US$60 million to municipalities investing in local public broadband networks, as well as low-interest loans to carriers to encourage them to build other broadband networks, including DSL, wireless, and cable systems.

1.3.6 South Korea

In February 2009, the Korea Communications Commission (KCC) announced a plan to upgrade the national network to offer 1 Gbps service by 2012. In wireless broadband aspect, the earlier plan was WiBro, which could offer seamless 100 Mbps hybrid networking. Its Heterogeneous Network Integration Solution (HNIS) technology combining 3G/4G service with any open Wi-Fi network to deliver speeds many times faster than North Americans can get from their wireless providers. The technology is designed to work without a lot of consumer intervention. HNIS will automatically provision open Wi-Fi access wherever the subscribers travel. The combination of mobile broadband with Wi-Fi works seamlessly as well. Currently, smartphones can use Wi-Fi or mobile data, but not both at the same time. While mobile operators cope with spectrum and capacity issues, HNIS can reduce the load on wireless networks without creating a hassle for wireless customers who used to register with every Wi-Fi service they encountered. The theoretical speed of an HNIS-enhanced 3G and Wi-Fi connection in South Korea will be 60 Mbps when SK Telecom fully deploys the technology. With 4G networks, theoretical maximum speeds are increased to 100 Mbps.

In 2014, the South Korea government launched the "Giga Korea Project" (GigaKorea) in both wired and wireless domains. The goal is to promote the growths of core technologies, future service, and innovative ecosystem so as to generate new jobs. According to the roadmap of GigaKorea, Korea will be able to provide hologram interaction service and supporting device and communication platform by 2020.

1.3.7 Singapore

The Singapore government plans to fund about US$520 million to establish the next-generation National Broadband Network. It covers wireline and wireless, and provides speeds ranging from 100 Mbps to 1 Gbps. The network will be opened to all service providers, instead of just limiting to major telcos such as Singtel and StarHub. However, the government will not prevent companies from building their own networks. Bidding consists of two stages. The first stage is the passive infrastructure. It has been started in September 2008. Information Development Authority (IDA) of Singapore selected OpenNet, which Singtel owns 30% shares for designing, building, and operating the passive infrastructure.

In the wireless area, IDA developed a wireless broadband programme, Wireless@SG, which was launched on December 1, 2006. Wireless@SG is a part of its Next-Generation National Infocomm Infrastructure initiative. In phase 1, Wireless@SG was operated by three wireless operators, iCell, M1, and Singtel. It was provided free to all Singapore residents and visitors. The phase 1 was completed in 2013. Users can enjoy free, both indoor and outdoor, seamless wireless broadband access with speeds of up to 1 Mbps at most public areas. Now, Wireless@SG programme is extended to phase 2 providing higher connection speed up to 2 Mbps and much more hotspots. So far, there are more than 2600 Wireless@SG hotspots for free WiFi access at many public places, such as restaurants, government offices, airport terminals, shopping malls, community clubs, hospitals, libraries, tourist attractions, and schools all over Singapore. Currently, these hotspots are operated and maintained by Singtel, StarHub, M1, and Y5ZONE.

Besides Wireless@SG, IDA in 2015 initiated a Heterogeneous Network (HetNet) programme in Jurong Lake District (JLD) as part of the Smart Nation initiative to align with the "Infocomm Media Masterplan 2025" announced by the Ministry of Communications and Information in Singapore. This programme is to deploy and test the advanced broadband wireless networks combining cellular mobile system and Wi-Fi to provide the customers seamless and much higher speed connection than either cellular or Wi-Fi systems.

1.3.8 Australia

In 2009, the Australian Government announced to establish National Broadband Network (NBN) through starting up a new company that will invest up to $43 billion over 8 years. The new network will provide optic fiber to home and workplace, as well as the next-generation wireless and satellite technologies to deliver superfast broadband services. In 2009, the government also released a discussion paper entitled "National Broadband Network: Regulatory Reform for 21st Century Broadband." The paper is based on public comments on the NBN. The paper appears to be similar to an FCC NPRM or NOI. It outlines the method of establishing the NBN and also sketches general regulatory reforms to assist the market in the future. To facilitate fiber build-out, the government will simplify land right of way procedures.

1.4 COGNITIVE RADIO

We could see from the various national broadband plans that more spectrums are required in order to fulfill the ever-increasing needs of wireless communications. We also understand from spectrum surveys conducted all over the world that there is not much spectrum available for allocation. On the contrary, the utilization of these allocated spectrums is low. This calls for a need to use spectrum in a more innovative manner.

The main regulatory bodies in the world such as FCC, Ofcom, and IDA as well as some independent measurements [3,7] found that most radio frequency spectrum was inefficiently utilized, except cellular network bands, which are overloaded in most areas of the world. A number of studies concluded that spectrum utilization depends on time and place. Moreover, these studies also show that the traditional fixed spectrum allocation prevents rarely assigned frequencies for specific services from being used, even when the unlicensed users do not cause noticeable interference to the assigned ones.

The above inefficiencies stimulated a new concept of Cognitive Radio (CR) where the radio itself is intelligent, aware of the environments, and able to learn and adjust its operating parameters in order to maximize delivery of wireless communication services. The concept of CR was first raised by Joseph Mitola III [8]. A CR is an intelligent radio that can detect the environmental conditions, including the spectrum availability, interference, and so on in its working area, and further dynamically configure its transmitter and receiver parameters so as to use the best wireless channel in the best way. This kind of mechanism can maximize the transmission of a user's own information, while also avoid the interference to the other spectrum users.

Some people equate CR as the extension of another technology called software defined radio (SDR). This is a misconception. SDR is a platform technology that allows the waveforms to be changed dynamically depending on needs since most processing is done at the digital domain. SDR may not need to understand the operating parameters of other users. SDR is being used even in cellular base stations so that changes to the system could be done quickly without having to replace the hardware in the deck. On the other hand, CR needs to understand the operating parameters of other users operating in the same environment since CR is typically operating as secondary users (SUs) who do not have the first right to the channels. CR could leverage on SDR platform for added flexibility but using SDR platform is not a prerequisite for CR.

Dynamic spectrum access (DSA) is one type of CR system. The major parameter that DSA needs to detect is frequency availability, which indicates the other users' spectrum utilization in the same working area. The other parameters also include the noise level, multipath, and so on. What a DSA system can reconfigure includes one to all of parameters consisting of operating frequency, signal waveform, bandwidth, emission power, antenna beam forming, network protocol, and topology.

Of the various features of DSA, the most challenging work technically is spectrum sensing, which detects the spectrum availability, namely, the other users' transmissions and frequencies. To protect the primary users (PUs), which are originally assigned to the spectrum, the spectrum sensing needs to be reliable even in multipath fading, blocking, and weak signal conditions (many decibels below noise level). This is the most challenging task. Most research work focuses on spectrum sensing.

It has been shown that a simple energy detector cannot guarantee the accurate detection of signal presence. To enhance the detection reliability, cooperative sensing that involves multiple nodes at different locations and sensing information exchange are necessary. This raises another issue since communications among these nodes are required before they could establish the communication links. This could be done either through in-band signaling or through a supplementary channel, for example, wireline Internet connection.

Besides spectrum sensing used in the CR system, a practical solution to protecting the primary users is to use a centralized database, which records the PUs' action details. Through checking the database, the SUs can avoid jamming the PUs. This is what Television White Space (TVWS) system is being designed currently. We will introduce TVWS in the next section.

1.5 TELEVISION WHITE SPACE

As mentioned earlier in this chapter, the usage of license–exempt spectrum grows in the past decade. Now many countries are taking the next step in the evolution of spectrum policy, through the license–exempt use of the TVWS. TVWS is the unused TV channels assigned to TV broadcasting. As digital TV has higher spectrum efficiency than analog TV, the worldwide trends of switching from analog to digital TV have freed up more spectrums in TV bands for TVWS. Spectrum sharing through TVWS is an important topic as it is the first step toward efficient use of spectrum in an opportunistic and dynamic manner. TVWS is expected to be the first CR system.

TVWS has superior propagation characteristics as it is in the very-high frequency (VHF) and ultrahigh frequency (UHF) spectrum bands. This results in longer communication distance and better penetration through obstacles. The long communication distance makes TVWS an attractive option for rural connectivity, while the better penetration gives TVWS an edge over other technologies for machine-to-machine applications in dense areas. Due to urbanization, TVWS provides a good option as the connectivity backbone for smart cities.

The favorable propagation characteristics of the TV white spaces promise to make white space devices (WSDs) even more powerful than their Wi-Fi ancestors. White spaces technologies will greatly expand the utility and help reduce the cost of using license–exempt devices to be used on broadband networks. It is also likely to make deployment of last-mile connections in hard-to-serve areas more affordable. These benefits, in turn, have significant economic and development benefits. Even though many of the world's citizens lack reliable Internet access, the Internet in 2010 has already contributed an average of 1.9% or $366 billion to GDP in 30 emerging markets.[1] With greater access to the Internet and to affordable devices, this number is certain to grow.

[1]O. Nottebohm, J. Manyika, J. Bughin, M. Chui, A.-R. Syed, Online and Upcoming: The Internet's Impact on Aspiring Countries, McKinsey & Company, 2012. Available at http://www.mckinsey.com/client_service/high_tech/latest_thinking/impact_of_the_internet_on_aspiring_countries.

White spaces access can be deployed quickly and takes advantage of otherwise unused spectrum. Because enabling spectrum access to white space devices does not require relocating incumbents and because the rules protect those incumbents, a license–exempt framework for access to TV white spaces can be adopted and put into use without disruption to incumbent operations.

The opportunistic or dynamic use concept central to license–exempt white spaces use ensures that rules can accommodate changing circumstances. Dynamic use is the idea that radio technologies should identify and use different frequencies within a defined band, based on what frequency is available for interference-free operation at a given time in a given geographic location. Although the particular unused TV channels vary from location to location, white space devices have the flexibility and agility to locate and operate on the unused channels, no matter where the devices are located in a country that permits such access. This means that previously unused spectrum becomes a valuable resource. It also means that both the technology and the rules can operate before, during, and after the digital television transition—regulators merely have to provide industry with information regarding occupied channels, switchover timeline, and new channel assignments and locations, and devices will be able to avoid broadcast operations and other licensed uses.

To enable more innovative services, the FCC released a notice of proposed rulemaking (NPRM) in May 2004 to allow unlicensed radio transmitters to operate in the broadcast TV spectrum at locations where that spectrum is not being used [9]. As TV band is the target spectrum to be used, this kind of system is also generally referred to as TVWS. Although CR concept could be used in any frequency bands, TVWS is gaining more traction due to the better propagation characteristics of TV bands that travel further and penetrate walls easier. In addition, the migration from analog to digital TV opens up more bandwidth for such innovation.

In order for white space devices to utilize TVWS, they must not cause noticeable interference to PUs. There are many proposed techniques to protect PUs from WSDs, such as spectrum sensing, beacon, and white space database (WSDB). While each technique has its own pros and cons, there was no consensus previously on which was the most suitable technique for determining TVWS channels.

To understand which technology is the most suitable for TVWS, the FCC conducted its first test of TVWS in 2007. The test concluded that spectrum sensing-based approaches had not reached the acceptable accuracy and reliability on detecting PUs. In 2008, FCC released another report after conducting the second trial, which concluded that the TVWS prototypes submitted by Adaptrum, Institute for Infocomm Research (I²R), Motorola, and Philips had met the burden of "proof of concept" in their ability to detect and avoid PUs' transmissions.

However, after receiving arguments from various parties, FCC finally decided that the current spectrum sensing technologies are not reliable enough to be used solely as a means to determine vacant TVWS. In 2010, FCC released a Memorandum Opinion and Order that determined the rules for TVWS that are fine-tuned several times subsequently with WSDB as the main method for determining vacant TVWS with spectrum sensing as an option. After the United States, Singapore also finalized its TVWS regulation in 2014, followed by the United Kingdom and

Canada in 2015. Besides the United States, Singapore, the United Kingdom, and Canada, CEPT, Finland, New Zealand, and other countries have also released their draft standards or regulatory frameworks for TVWS. WSDB remains as the main choice for determining vacant TVWS channels.

TVWS becomes the first spectrum open to CR for real application for better spectrum utilization. However, one should not be confused between TVWS, being a new way of accessing frequency spectrum, and specific wireless products such as Wi-Fi. For illustration, the land transport regulator could define the usage of lanes in a particular road to be normal lanes, carpool lanes, and so on. When there is emergency vehicle on the way, normal cars have to give way. The lanes are analogy to TVWS, while the cars are analogy to Wi-Fi and other wireless technologies. This also means that Wi-Fi could operate in TVWS spectrum and so is cellular.

1.5.1 TVWS Regulation

Although TVWS applications rely mainly on WSDB, there are other requirements such as protection contour, self-positioning, out-of-band (OOB) limits, update cycle, and so on that are important parameters to be aware of when implementing TVWS in specific countries as these parameters differ for different territories. Thus, suitable regulation is necessary.

In the regulatory front, FCC is the clear leader as it was the first regulator to experiment TVWS in a series of prototype trials [10,11]. In November 2008, FCC announced opening up of TVWS for unlicensed access after it was convinced that the prototype it tested met its initial expectations [12]. This marked a critical milestone in TVWS development worldwide as this groundbreaking regulation is often being referred to in subsequent regulations carried out in other countries. The initial FCC regulation was later superseded with updates in the final rules and regulation for the use of TVWS in 2010 [13]. After considering further comments to the regulation, the rules were further updated in April 2012 [14]. The release of the third memorandum can be interpreted as stabilization of TVWS regulation since most of the concerns raised have been resolved. In 2014, FCC released a new notice proposing some amended rules and operation parameters, although it has not been finalized yet. The proposed changes included allowing white space device operations in some reserved channels for wireless microphones and a few prohibited channels before with some conditions such as requiring WSDs to access WSDB more frequently and emission power restrictions. The detailed information can be found in Chapter 2.

After the successful FCC regulation, the Office of Communications (Ofcom) in the United Kingdom followed suit but adopting a different approach. In the United States, various trials were carried out by individual or groups of organizations without much coordination. In contrast, a strong consortium was formed in the United Kingdom called the Cambridge White Spaces Consortium to carry out various trials in Cambridge with the blessing from Ofcom. Satisfied with the outcome, Ofcom has also approved license–exempt use of TVWS in the United Kingdom in

2011 [15]. In other parts of Europe, the Electronic Communication Committee (ECC) is also actively looking into adopting TVWS where the Finnish Communications Regulatory Authority has issued a test license to implement Europe's first TV white space geolocation database in August 2012.

In Asia, Singapore is the leader in TVWS development where the Cognitive Radio Venue (CRAVE) trial was conducted in March 2011 [16]. Subsequently, the Singapore White Spaces Pilot Group (SWSPG) was formed in April 2012 by the Institute for Infocomm Research, Microsoft, and StarHub to look at commercial pilots of TVWS technologies [17]. The SWSPG is unique compared with its American and European counterparts in the following aspects:

- The test environment is different where Singapore is a city state with many high-rise buildings.
- Singapore is "sandwiched" between Malaysia and Indonesia. Many of the interleaved TV bands, as a result of coordination with Malaysia and Indonesia, could be exploited provided the pilots do not interfere with Malaysia and Indonesia users.
- SWSPG involves many end users apart from technology providers. This will bring TVWS to the next level of potential commercialization.

Apart from the few countries mentioned above, there are also a growing number of developments and trials conducted elsewhere worldwide as summarized in Chapter 5. This further confirms the momentum of TVWS worldwide.

1.5.2 Standardization

To respond to the new FCC initiative, IEEE started a new standardization activity to look at how to reuse TV broadcast bands as secondary access for Regional Area Network under the IEEE 802.22 standard [18]. The main application of IEEE 802.22 is to serve rural areas using the range benefits of TVWS. In July 2011, this first IEEE standard on TVWS was released to mark an important milestone in the history of TVWS standardization.

Following this, the IEEE 802.11 standardization also tried to leverage on this new method of spectrum access to come out with IEEE 802.11af standard in order to allow Wi-Fi to operate in the TV bands. This will allow TVWS to be used as Wi-Fi hotspot or other applications.

Another IEEE standardization group for Wireless Personal Area Network (WPAN) also proposed to use TVWS as IEEE 802.15.4m standard. This will allow WPAN to extend its range so as to serve WPAN as well as machine-to-machine (M2M) application. Another proprietary standard weightless is also looking into the use of TVWS for M2M.

Some might be confused by the many systems defined by different standards as they view TVWS as a single system. In reality, TVWS is a method of gaining access to spectrum. After the spectrum is secured, one can design many different systems using this spectrum. An easier to understand example is the ISM band.

Once the ISM bands are allocated, many different systems can be designed using the same bands, which include Wi-Fi, Bluetooth, ZigBee, baby monitoring, drone control, and so on. The same applies to TVWS.

Given the current momentum, it is believed that TVWS will be an option for many standards that uses frequency spectrum. Reference [19] gave a summary of various standardization activities related to CR.

1.5.3 Potential Applications

Due to range and penetration benefits as well as the potentially huge frequency spectrum available, many applications are possible with TVWS.

Machine-to-Machine Communications (M2M) One of the hottest applications of TVWS is for M2M communications. Traditionally, M2M faced problem of range from the current ISM band solutions or cost from public network solution. TVWS is positioned uniquely between these two where it is license–exempt as well as have good range. One of the earliest M2M applications is smart grid.

Super Wi-Fi Globally, mobile data consumption rises exponentially. This is due to the increase in the number of mobile broadband subscribers as well as increased use of mobile broadband per user. This tremendous increase in mobile data has resulted in great pressure over the current cellular mobile network. Mobile offloading has been exploited to alleviate this problem and Wi-Fi is a leading candidate for this option. Nevertheless, due to its range limitation, Wi-Fi offloading will incur high capital expenditure for setting up of the network as well as high operating expenses for the lease line for the backbone. TVWS is suitable to reduce the burden of Wi-Fi due to its superior range and abundant frequency spectrum, namely, super Wi-Fi.

In-Home Distribution There is also an increase of data going into the home. However, in-home distribution still lacks convincing solution. While Broadband Powerline Communication (BPLC) could be a possible solution, the use of BPLC for large-scale adoption has not been proven. In addition, the bandwidth allocated for BPLC is also limited. Therefore, it will limit the future expansion of BPLC technology. TVWS could be a good candidate for in-home distribution given its good penetration characteristics.

Video Surveillance Another promising area for TVWS is to use it for wireless video surveillance. Until now, wired solution is still preferred for video surveillance primarily due to the lack of bandwidth for the current wireless solutions. However, wired solution is expensive due to setup cost. This will slow down video surveillance deployment. The current popular wireless solution for video surveillance is to use 3G data connection. However, one should bear in mind that video surveillance is using the uplink channel that has much lesser bandwidth than the downlink in 3G. Due to this, the current 3G-based wireless solution is not able to scale up quickly. To enable fast wireless video surveillance deployment, TVWS is a good candidate. It provides the necessary data rates to support high-quality video surveillance with the abundant bandwidth. It also gives the video surveillance operators freedom to deploy surveillance cameras at their preferred locations.

Disaster Planning   In 2015, Gigabit Libraries Network and State Library Agencies in the United States started to explore the possibility of using the portable TVWS broadband equipment in community disaster planning. The rationale is that after a disaster, the public communications infrastructure could be down for 2–5 days. In that case, the community needs temporary connectivity until the public utilities recover. A TVWS network can help fill the communications gap through deploying temporary Internet hotspots around the community in much greater distances compared with that the traditional Wi-Fi can reach. The TVWS trial for disaster planning will be partially funded by a grant from the Knight News Challenge on Libraries. It will explore how libraries can be a platform to build more knowledgeable communities.

Apart from the applications mentioned above, there are many other possible applications such as telemetry, neighborhood network, transportation, maritime, and so on. In short, the potential application can only be limited by our imagination.

1.5.4 Technologies

The original idea of CR and TVWS was to use technological approach to identify underutilized frequency spectrum for communication. The key technology to achieve this purpose was to use spectrum sensing to detect primary users (e.g., broadcast signals and wireless microphone signals). Spectrum sensing could be classified into blind sensing and feature sensing. The former does not need to know the waveform of the intended signal to sense, whereas the latter needs to know certain features, although not all, of the primary user. To improve sensitivity of spectrum sensing, coordinated sensing may be adopted. Reference [9] provides some comparisons of the different sensing schemes.

In the trial conducted by FCC in 2008, spectrum sensing was the main focus, although the geolocation database (WSDB) from Motorola was also tested [13]. As a result of concerns from the primary users in the confidence level of spectrum sensing accuracy, WSDB approach emerged as the preferred option for accessing TVWS. This can be seen from both FCC and Ofcom's regulatory framework. For wireless microphone, a safe harbor approach is adopted.

WSDB is more a policy solution rather than a technological solution, although certain technologies are involved. Before accessing TVWS frequency spectrum, a TVWS device sends a query to a WSDB server stating its current location. The WSDB server then responds with the list of frequency channels available for communication.

The WSDB server obtains the information of primary users through regulators as well as user inputs. In the United States, FCC awarded licenses to 10 organizations to operate WSDB. Currently, the WSDB from Spectrum Bridge, iconetiv, Google, RadioSoft, LSTelcom, and Key Bridge Global have been approved for operation in the United States.[2]

Although WSDB is the *de facto* option for TVWS spectrum access now, FCC does not rule out spectrum sensing only devices. In fact, in countries where regulation executions are weak, technological solution such as spectrum sensing might be preferred. For CR systems other than TVWS, WSDB implementation might not be easy if the primary usage pattern can only be determined in an *ad hoc* manner. Therefore, spectrum sensing is still a hot topic in the research community.

1.5.5 Moving Forward

Although TVWS has some promising applications and results now, the system can be further enhanced when it becomes more mature in the near future. This section touches briefly three aspects of potential enhancements to the current TVWS systems. More details of the future directions can be found in Chapter 7.

Emission Masks   One of the most challenging requirements for TVWS is the emission masks requirements, especially the emission masks specified by the FCC. One of the reasons for stricter emission mask requirements from FCC was due to technology uncertainty when FCC opened up TVWS previously. Other non-technical reason included protest from the existing primary users on the potential interference with their existing system. As there were not enough data to back these claims, FCC adopted a more cautious approach with regard to the emission mask requirements. With more studies conducted and results collected, the requirements on emission masks were relaxed as can be seen from the regulation by Ofcom [20] and Singapore [21]. Moving forward, there is no reason why FCC could not lower its emission mask requirements if the other regulators and trial results show its viability.

WSDB with Sensing   There is almost consensus of using WSDB for TVWS access currently. The present WSDB is conservative in the sense that a large range of protection contour is defined in order to ensure no interference to the primary users, although the primary user might be far apart. This results in many otherwise vacant channels being left unused. This also presents another opportunity to further enhance spectrum occupancy rates.

One approach is to combine WSDB with spectrum sensing. By using real-time spectrum sensing results provided by TVWS devices or other devices operated to gather channel information, the protection contour could be adjusted according to the real information gathered in the field. Spectrum sensing results will also help in

---

[2] https://www.fcc.gov/encyclopedia/white-space-database-administrators-guide.

fine-tuning the various parameters used in WSDB calculations such as terrain profile, climate profile, and so on. On the contrary, WSDB information could also be used to adjust the sensitivity (and thus threshold) requirements for spectrum sensing so as to increase the results of spectrum sensing. In short, both technologies could coexist and reinforce each other in order to ensure better spectrum utilization. With this approach, spectrum sensing only system will have the opportunity to emerge in the future when the end users have better confidence in spectrum sensing.

WSDB with Quality-of-Service Support   The current WSDB only protects the primary users. It does not guarantee the performance of the secondary users (SUs). While this is understood from the current regulation practices, it has to be enhanced so as to allow differentiated access even for SUs. Otherwise, quality of service (QoS) may become a big issue when TVWS is widely adopted.

A novel method called WSDB with QoS (WSDB-Q) can grant access to higher priority communication via reserved channels based on a set of criteria predefined in order to maintain QoS of communication systems accessing TVWS spectrum. The set of criteria covers regulatory, technical, and commercial aspects. The details of WSDB-Q will be presented in Chapter 3.

1.5.6   Features of TVWS

TVWS are using the frequency bands originally assigned to TV broadcasting in an opportunistic manner. The signals in these frequency bands have strong propagation and penetration. A single broadcast TV tower can cover a large area with sufficient quality of reception. Compared with the Wi-Fi frequencies and most cellular systems, a TVWS signal can reach longer transmission distance with the same transmission power, or emit lower transmission power for the same communication distance. Moreover, TVWS span a large frequency range: from VHF to UHF. The potential TVWS bandwidth is more than 500 MHz. This is much wider than that currently assigned to cellular systems. This means huge opportunities for TVWS even if only a fraction of it is being explored currently.

1.6   SUMMARY

In this chapter, we showed some spectrum survey results and understood that the current spectrum utilization is very low, although radio spectrum is almost fully allocated. This shows that the traditional spectrum allocation is not sustainable. New dynamic spectrum sharing is necessary to meet the requirements of the wireless explosion era.

We introduced CR, the new technology that can sense the radio environment before using the spectrum with a suitable transmission approach. This leads to much higher spectrum utilization.

TVWS is the first CR system for real applications. As of now, most of the major authorities have crafted TVWS regulations to support TVWS development as well as protect the existing spectrum users. There are significant potentials with wide application areas. Also, there are challenges in terms of technology and regulation and commercialization. TVWS is a pioneering system for DSA. The experiences gained from TVWS can be used in other frequency bands adopting DSA. Thus, studying TVWS is essential for the future new paradigm of spectrum allocation and utilization.

In this book, we discuss TVWS in terms of regulation, technology, standardization, deployment, and commercial and market potential as well as development beyond TVWS. In Chapter 2, we will discuss various TVWS regulations across the world and attempt to provide some comparisons so that the readers can understand the rationale and considerations behind different regulations. We will explain the technical challenges in TVWS and some novel solutions in Chapter 3. The standardizations related to TVWS will be described in Chapter 4. In Chapter 5, the status of global TVWS deployments will be discussed. The commercial and market potentials of TVWS are explored in Chapter 6. Finally, we will list some further development topics related to TVWS.

REFERENCES

1. https://www.ntia.doc.gov/files/ntia/publications/january_2016_spectrum_wall_chart.pdf.
2. National Radio Astronomy Observatory (NRAO), Spectrum Occupancy Measurements Location 5 of 6, Revision 3, Green Bank, West Virginia, October 10–11, 2004.
3. Md. H. Islam, C. L. Koh, S. W. Oh, X. Qing, Y. Y. Lai, C. W., Y.-C. Liang, B. E. Toh, F. Chin, G. L. Tan, and W. Toh, Spectrum survey in Singapore: occupancy measurements and analyses. *Cognitive Radio Oriented Wireless Networks and Communications*, IEEE, 2008.
4. J. Eggerton, FCC broadband plan: Commission sets 2015 spectrum deadline, Broadcasting & Cable, March 15, 2010 (retrieved March 23, 2010).
5. FCC, National Broadband Plan, Broadband Action Agenda, March 17, 2010.
6. Digital Britain Implementation Plan, August 2009.
7. V. Valenta et al., Survey on spectrum utilization in Europe: measurements, analyses and observations, in Proceedings of the Fifth International Conference on Cognitive Radio Oriented Wireless Networks & Communications (CROWNCOM), 2010.
8. J. Mitola, III, and G. Q. Maguire, Jr., Cognitive radio: making software radios more personal. *IEEE Pers. Commun.*, 6, 13–18 (1999).
9. Federal Communications Commission, Notice of Proposed Rule Making FCC 04–113: Unlicensed Operation in the TV Broadcast Bands (ET Docket No. 04–186, FCC), 2004.
10. S. W. Oh, Y. Ma, M.-H. Tao, and E. C. Y. Peh, TVWS in the world: regulations and comparisons, in Proceedings of the International Conference on Frontiers of Communications, Networks and Applications (ICFCNA), 2014.
11. S. K. Jones, T. W. Philips, H. L. Van Tuyl, and R. D. Weller, Evaluation of the performance of prototype TV-band white space devices phase II, FCC, 2008.
12. Federal Communications Commission, Second Report and Order and Memorandum Opinion and Order, in the Matter of ET Docket No. 04–186 and ET Docket No. 02–380, FCC, 08–260, 2008.
13. Federal Communications Commission, Second Memorandum Opinion and Order in the Matter of Unlicensed Operation in the TV Broadcast Bands (ET Docket No. 04–186) and Additional Spectrum for Unlicensed Devices Below 900 MHz and in the 3 GHz Band (ET Docket No. 02–380), FCC, 2010.
14. Federal Communications Commission, Third Memorandum Opinion and Order in the Matter of Unlicensed Operation in the TV Broadcast Bands (ET Docket No. 04–186) and Additional Spectrum for Unlicensed Devices Below 900 MHz and in the 3 GHz Band (ET Docket No. 02–380), FCC, 2012.

15. Office of Communications, Implementing geolocation: summary of consultation responses and next steps, September 1, 2011.

16. Infocomm Development Authority of Singapore, Trial of white space technology: accessing VHF and UHF bands in Singapore, June 2010.

17. Singapore White Spaces Pilot Group, D 2015. Available at http://whitespace.i2r.a-star.edu.sg/ (accessed February 15, 2015).

18. IEEE 802.22 Wireless Regional Area Networks, D 2015. Available at http://www.ieee802.org/22/ (accessed February 15, 2015).

19. Y. Zeng, Y-C. Liang, Z. Lei, S. W. Oh, F. Chin and S. Sun, Worldwide regulatory and standardization activities for cognitive radio, C in 2010 IEEE Symposium on New Frontiers in Dynamic Spectrum, IEEE, Singapore, 2010.

20. W. Webb, On using white space spectrum, *IEEE Commun. Mag.*, V 50, 145–151 (2012).

21. IDA, Regulatory framework for TV white space operations in the VHF/UHF bands, June, 2014.

Chapter 2

Regulations

This chapter covers the regulation activities in TVWS. Regulation is important for TVWS since it is a new way of using spectrum that will affect the current licensed users. Regulations must be in place to define the operation parameters of WSDs such that it will ensure that the current licensed services and users are not being disrupted by WSDs. The International Telecommunication Union (ITU) has concluded that the current spectrum framework is sufficient to cover TVWS and it is up to the respective nations to come out with their regulations [1]. TVWS is evaluated by many regulators to fulfill the demand of radio spectrum for new and existing wireless services. However, regulation is a top–down process where typically more developed countries are able to handle better compared to developing countries. In particular, the regulators in the United States and the United Kingdom are in the forefront in establishing regulations for the operation of WSDs in the TV bands. Developing countries tend to rely on regulations from the developed countries. To help speed up regulation development for developing countries, the Dynamic Spectrum Alliance (DSA) is introducing a model rule (see Appendix 1) to help developing countries to jump start their regulation processes.

In North America, the FCC of the United States published the first TVWS regulation in the world. FCC has completed the revision of TVWS rules, and is pursuing dynamic spectrum access rules for other frequency bands such as 3.5 and 5 GHz. Canada largely follows the FCC.

In Europe, UK's Ofcom is the lead in terms of developing TVWS rules. ETSI TVWS rules have been driven from the UK rules. CEPT is also looking at other approaches like licensed shared access (LSA) and licensed assisted access (LAA).

In Asia Pacific, Singapore's IDA has the TVWS regulations ready. Singaporean TVWS players are keen to push their regulations to ASEAN countries. Philippines and Indonesia are already working toward TVWS regulations. India is the biggest developing market economy with significant interests in TVWS. China also has several ongoing TVWS activities, but not much detail is available publicly. The New Zealand Radio Spectrum Management (RSM) has also implemented TVWS devices certification and licensing rules.

TV White Space: The First Step Towards Better Utilization of Frequency Spectrum, First Edition.
Ser Wah Oh, Yugang Ma, Ming-Hung Tao, and Edward Peh.
© 2016 by The Institute of Electrical and Electronics Engineers, Inc. Published 2016 by John Wiley & Sons, Inc.

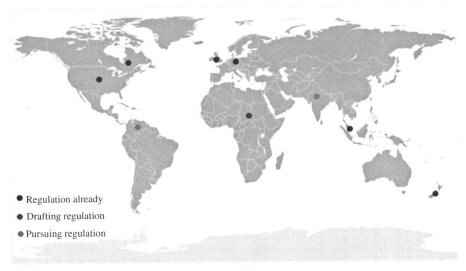

Figure 2.1   TVWS regulation status around the world.

In Africa, a number of countries are active in TVWS. Malawi is in the fore-front and could most likely be the first African country to enact TVWS regulations. Ghana is also about to start TVWS regulation process. Botswana aims to look at TVWS rules in 2016. Nigeria is the chair of African Telecommunications Union (ATU) TVWS Study Group.

In Latin America, although there are no TVWS rules expected in the short to medium term, Colombia, Brazil, and Mexico are starting to look at TVWS.

The challenges to enable dynamic reuse of licensed spectrum either spatially or temporally while ensuring disruption-free operations of licensed users are evaluated by many regulators. These challenges need to be addressed before regulators approve the use of WSDs in TVWS. Therefore, understanding the regulations for WSDs set by the regulators to meet these challenges are of great interest to both the researchers and developers of WSDs. The objective of this chapter is to provide review and comparison of the different TVWS regulations set by various countries, in particular, countries in North America, Europe, and Asia Pacific. The status of the TVWS regulation around the world is summarized in Figure 2.1.

2.1   NORTH AMERICA

2.1.1   The United States of America: FCC

In 2004, the Federal Communications Commission (FCC) of the United States became the first regulator in the world to propose the use of unlicensed devices to operate in TV bands at locations where the TV bands are not being used [2]. The purpose is to provide more efficient and effective use of TV bands, which may lead to developments of new services or increase the communication range of existing services. This is also consistent with the significant growth and demand for unlicensed wireless broadband services for which the unused TV bands could provide the additional spectrum for such applications. The proposal came at a time when TV broadcasters are still undergoing a transition from analog to digital transmissions, which should free up more TV spectrum when the transition is completed. The FCC also recognizes the needs to ensure the new unlicensed devices, which are also called white space devices (WSDs), will not cause harmful interference to incumbent TV band users such as TV broadcasters and wireless microphones. In FCC's first proposed rulemaking, several methods to prevent harmful interference to the incumbent users are being considered, such as (1) allowing current broadcasters to transmit information of available TV channels to unlicensed devices, (2) employing geolocation technologies, and (3) spectrum sensing to detect the incumbent transmitters.

Two years after the proposed rulemaking was published, the first report and order for unlicensed operation in TV bands was released in 2006 [3]. In this report, initial decisions are made to allow fixed low-power WSDs to operate at times and locations where the TV bands are not being used by the incumbent users, except for TV channels 2–4 and 14–20 that were still in consideration at that time. Unlicensed personal portable devices are being considered to operate in TV bands without causing harmful interference to the incumbent users, but are determined that they shall not be allowed to operate on TV channels 14–20, which are used by public safety services. This is because personal portable devices may pose a greater risk of harmful interference as it is hard to track their locations and their antennas may be less efficient to sense the incumbent signals. It is also decided that TV channel 37, which is used by radio astronomy and wireless medical telemetry services, should not be used by any unlicensed WSDs.

In the second report and order for unlicensed operation in TV bands released in 2008, many details for unlicensed operation in TVWS are determined [4]. Many of the TV bands that were being considered for unlicensed WSDs in the first report [3] have been approved. Fixed devices are allowed to operate in channels between 2 and 51 except channels 3, 4, and 37, while personal portable devices are allowed to operate in channels between 21 and 51 except channel 37. The transmit power limits of the unlicensed WSDs are also determined, whereby fixed devices are limited to 4 W effective isotropic radiated power (EIRP) and personal portable devices are limited to 100 mW EIRP. For adjacent channels of broadcast channels, personal portable devices are further limited to 40 mW EIRP, while fixed devices were not allowed to operate. The initial decision of out-of-band (OOB) emissions for unlicensed WSDs in this report were set to be 55 dB below their in-band power level as measured in 100 kHz bandwidth. It is also determined that unlicensed devices, except for personal portable devices operating in client mode, must have geolocation capability and shall be able to access the WSDB to obtain a list of available channels for their operations. At this stage, it was determined that all unlicensed devices must also have sensing capabilities of detecting incumbent signals at -114 dBm, in addition to WSDB access, to further minimize the interference to the incumbent users. Unlicensed WSDs that have only spectrum sensing capability are also allowed, but they must be subjected to much more rigorous tests by FCC before being used by public and also with a transmit power limit of 50 mW ERIP. At this stage, it was also determined that two of the TV channels will be free from unlicensed WSDs so as to ensure wireless microphone users have available channels to use. Wireless microphones can also preregister their usages in the WSDB for protection against harmful interference from unlicensed WSDs by having a separation distance of at least 1 km between them.

In the report, a method has been defined to determine whether a TV channel is occupied by the TV broadcaster at a particular location. This method computes a contour around TV transmitters and the area within the contour is considered to be occupied by TV receivers and shall be protected from the unlicensed WSDs. The signal levels of the TV transmitters will define the protection contours and the method to compute the signal levels depends on the type of TV signals. The signal levels are computed using the $F(50,50)$ curves specified in FCC rule in radio broadcast services for analog TV signals and the $F(50,90)$ curves for digital TV signals [5]. $F(x,y)$ means the predicted electromagnetic field strength occurs at $x\%$ of the receivers $y\%$ of the time. The unlicensed WSDs were required to ensure that they operate outside the protection contour with a sufficient separation distance to ensure that the desired-to-undesired signal ratios of TV receivers within the protection contour are above a threshold. An example of an available TV channel location for unlicensed WSDs based on the protection contours and separation distance method is illustrated in Figure 2.2.

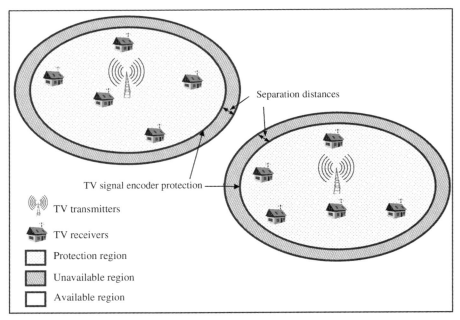

Figure 2.2   Available TV channel locations for unlicensed devices based on the protection contours and separation distance.

In the same report, some rules are set for database administrators. It is stated that database administrators are permitted to charge fees from unlicensed WSDs for the service of providing a list of available channels to them. The WSDB are required to contain information on the following:

1. All authorized services using fixed transmitters with designated service areas that operate in TV bands
2. Service paths of broadcast auxiliary point-to-point facilities
3. Operation locations of private land radio service and commercial mobile radio service in channels 14–20
4. Operation locations of offshore radiotelephone service
5. Locations of cable head ends and low-power TV receive sites that are outside of protection contours of TV transmitters
6. Locations and schedules of registered wireless microphones and other low-power auxiliary devices

In 2010, the FCC published a second memorandum opinion and order to finalize and modify the rules for unlicensed operation in TVWS [6]. One of the major changes from the second report in 2008 is that unlicensed WSDs that use WSDB access to determine the available TV channels are no longer mandatory to include spectrum-sensing capability. For sensing-only personal portable devices, the detection threshold to detect wireless microphones is relaxed to −107 from −114 dBm. In this report, power spectral density (PSD) limits for unlicensed WSDs are introduced to prevent unlicensed WSDs that have bandwidths of significantly less than 6 MHz to have high power concentration in narrowband. This is because if multiple unlicensed WSDs have high power concentration in narrow bands across the same TV band, it will result in a total transmit power significantly higher than the power limits of a single unlicensed WSD that operates over the entire 6 MHz. The rule for OOB emission was revised to 72.85 dB below their in-band power level as measured in 100 kHz bandwidth. However, this OOB rule is temporary as it is changed again in the third memorandum opinion and order [7].

More details are also provided on the operations of unlicensed personal portable devices in this report that are summarized here. The unlicensed personal portable devices are subcategorized into mode I, mode II, and sensing-only devices. The main difference between mode I and mode II devices is that mode II devices have the capability to determine the available channels at their locations, while mode I devices do not have that capability and need to be controlled by fixed or mode II devices. Mode I devices need to receive signals from fixed or mode II devices at least once every minute to ensure the availability of the channel. For mode II devices, they need to recheck their locations at least once every minute and have to redetermine the available channels if they move more than 100 m. Mode II devices need to access WSDB at least once a day to determine the availability of the TV bands even if they have not moved more than 100 m. The separation distance between mode II devices and registered wireless microphone users is revised from 1 km to 400 m when operating in co-channel.

The report also decided that the two channels that were reserved for wireless microphones will be the first vacant channel above and below channel 37. A new rule is also introduced to allow operators of event and production shows that use large number of unlicensed wireless microphones to register for the same protection as licensed wireless microphones against unlicensed WSDs.

For database operations, new rules include security features for online financial transactions and confidential information, rejecting queries of available channels from uncertified unlicensed WSDs, and methods to prevent data corruption and unauthorized data modification are also introduced. The WSDB must provide channel availability information that includes schedule changes over the course of 48 h period from the time an unlicensed WSD makes the query. Some additional features are optional for WSDB such as "push" system to update the unlicensed devices of any information change and ranking the quality of channels for unlicensed WSDs.

After the second memorandum opinion and order to finalize rules for unlicensed operation in TVWS in 2010, the FCC released a third memorandum opinion and order in 2012 to modify three rules [7]: (1) increasing the maximum height above average terrain (HAAT) for sites where fixed devices may operate, (2) changing the OOB emission limits to specify fixed values instead of relative to in-band power value, and (3) increasing the maximum PSD values slightly. The maximum permissible HAAT for fixed devices, which was decided to be 76 m HAAT plus 30 m maximum antenna height or 106 m in second memorandum opinion and order, is now revised to 220 m HAAT plus 30 m maximum antenna height or a total of 250 m. The reason given is that when constraining fixed devices to sites with HAAT that is no greater than 76 m will preclude the operation of fixed devices in many locations, particularly in rural areas. The reason for setting the OOB powers to fixed values is that devices that transmit at less than maximum power will no longer need to suppress their OOB powers to values lesser than devices that transmit at maximum power. This also simplifies compliance measurements, equipment designs, and hence reducing cost. The reason to increase the PSD slightly is to allow the unlicensed WSDs to operate at maximum power at a bandwidth of 5.5 MHz with 250 kHz for roll-off from in-band signal to each adjacent channel.

All the latest operation parameters of unlicensed devices in TVWS determined by FCC up to the third memorandum opinion and order are provided in Table 2.1.

The minimum separation distances that WSDs must be located from the protection contours of co-channel and adjacent channel TV transmitters are shown in Table 2.2. The signal levels that define a TV transmitter's protection contour for both analog and digital TV stations are shown in Table 2.3.

However, in the latest development of incentive auction in the United States, TV broadcasters will receive financial compensation if they allow some of the spectrum to be auctioned for licensed wireless services. This may affect TV spectrum available for unlicensed WSDs. In addition, there will be new licensed services in the 600 MHz band that the unlicensed WSDs have to protect as well. Hence, FCC has released a new notice of proposed rulemaking to amend the rules for unlicensed operations in TV bands in 2014 [8]. The new proposed rules and operation

Table 2.1 Operating Parameters of WSDs under Regulation of FCC

Devices	Fixed Devices	Mode II	Mode I	Sensing-Only
Transmit power limit (EIRP)	4 W; not allowed for channels adjacent to incumbent	100 mW; 40 mW for channels adjacent to incumbent	100 mW; 40 mW for channels adjacent to incumbent	50 mW
In-band channel PSD limit (100 kHz)	12.6 dBm; not allowed for channels adjacent to incumbent	2.6 dBm; −1.4 dBm for channels adjacent to incumbent	2.6 dBm; −1.4 dBm for channels adjacent to incumbent	−0.4 dBm
Adjacent channel PSD limit (100 kHz)	−42.8 dBm	−52.8 dBm; −56.8 dBm for channels adjacent to incumbent	−52.8 dBm; −56.8 dBm for channels adjacent to incumbent	−55.8 dBm
Permissible TV channels	2, 5–36, 38–51	21–36, 38–51	21–36, 38–51	21–36, 38–51
Permissible TV frequencies (MHz)	54–60, 76–88, 174–216, 470–608, 614–698	512–608, 614–698	512–608, 614–698	512–608, 614–698
Discover available channel	WSDB access	WSDB access	From fixed device or mode II device	Spectrum sensing
Maximum time/distance to redetermine channel availability	Once a day	Once a day or move more than 100 m	Once every minute	Once every minute
Geolocation capability	Accuracy ± 50 m	Accuracy ± 50 m, recheck location once every minute	NA	NA
Sensing capability	NA	NA	NA	Digital TV: −114 dBm, (6 MHz), analog TV: −114 dBm, (100 kHz), wireless microphone: −107 dBm, (200 kHz)
Separation distance from wireless microphones	1 km	400 m	400 m	NA

29

Table 2.2   Separation Distances of WSDs from Protection Contour

Antenna Height above Average Terrain of Unlicensed Device (meters)	Required Separation from Digital or Analog TV Protection Contour	
	Co-channel (km)	Adjacent Channel (km)
Height < 3	4.0	0.4
3 ≤ height < 10	7.3	0.7
10 ≤ height < 30	11.1	1.2
30 ≤ height < 50	14.3	1.8
50 ≤ height < 75	18.0	2.0
75 ≤ height < 100	21.1	2.1
100 ≤ height < 150	25.3	2.2
150 ≤ height < 200	28.5	2.3
200 ≤ height ≤ 250	31.2	2.4

parameters in the proposed rulemaking will be highlighted in this section. However, at the time of writing this book, these new proposed rules and operation parameters are still subject to changes after feedbacks from the relevant stakeholders and therefore are not finalized by FCC.

The proposed changes to the rules for TVWS operation include allowing operation of unlicensed WSDs on the first two vacant TV channels above and below channel 37, which are previously reserved for wireless microphones. Hence, to protect the wireless microphones, it also proposes to decrease the time interval, which unlicensed WSDs access WSDB to verify any changes in channel availability, to 20 min so that an unlicensed WSDs will cease operation within 30 min when a new microphone registration is made in a WSDB. Registration of unlicensed wireless microphones in WSDB is no longer permitted. The new proposed rulemaking is considering removing the prohibition on the use of channels 3 and 4 by fixed unlicensed devices as the interference to TV interface devices and TV receivers in channels 3 and 4 may be limited. It is also considering removing the prohibition on the use of channels 14–20 by unlicensed personal portable devices as the WSDB is reliable enough to protect private land radio service and commercial mobile radio service from personal portable devices in channels 14–20. The commission is also

Table 2.3   Minimum TV Signal Levels within Protection Contour in the United States

Type of Station	Protected Contour		
	Channel	Contour (dBμ)	Propagation Model
Analog: class A TV, low-power TV, translator, and booster	Low VHF (2–6)	47	F(50,50)
	High VHF (7–13)	56	F(50,50)
	UHF (14–69)	64	F(50,50)
Digital: full-service TV, class A TV, low-power TV, translator, and booster	Low VHF (2–6)	28	F(50,90)
	High VHF (7–13)	36	F(50,90)
	UHF (14–69)	41	F(50,90)

considering the operation of unlicensed devices in channel 37 with some limitations. For example, fixed devices are able to operate at 4 W in channel 37 when both channels 36 and 37 are also vacant and 40 mW when both channels 36 and 37 are occupied. If one of the adjacent channels is vacant, the fixed device can still operate at 4 W, but its center frequency must be at the boundaries of channel 37 and the vacant channel.

There are proposed changes to the transmit power operations of unlicensed WSDs in the new proposed rulemaking. First, fixed devices are allowed to operate in channel adjacent to occupied TV channels as long as their transmit power is lower than 40 mW, which is the same as personal portable devices. Hence, a co-channel and adjacent separation distances for fixed device at 40 mW and intermediate power levels up to 4 W are provided in the proposed rulemaking to protect other wireless services, including services in channel 37 and the repurposed 600 MHz band. The PSD limit per 100 kHz for fixed devices is provided at 4 W EIRP in the third memorandum opinion and order; however, in this rulemaking, the PSD for intermediate power levels from 40 mW to 4 W are provided. The commission is also considering to increase the transmit power of unlicensed WSDs in rural areas but no decision has been made yet. A new rule is proposed to enable fixed devices with maximum power of 4 W to operate at locations where there are two contiguous available channels instead of three, which is illustrated in Figure 2.3. To enable channel bonding, the OOB emission limits will not be applied when the adjacent channel is being used by the same unlicensed device. The OOB limits will only be applied to the 6 MHz bands immediately above and below the edges of the band of contiguous channels used by the unlicensed device.

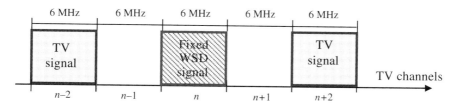

(a) Fixed WSD can operate at maximum power when there are three contiguous vacant channels under the rules in third memorandum opinion and order.

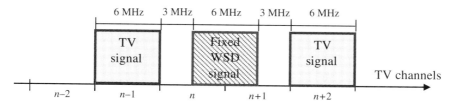

(b) Fixed WSD can operate at maximum power when there are two contiguous vacant channels under the new proposed rulemaking.

Figure 2.3 Permissible operation channel of fixed WSDs under the new proposed rulemaking.

Table 2.4 New Proposed Changes to Operation Parameters of Fixed Device

Device Type	Fixed Device
Transmit power limit (EIRP)	4 W, 40 mW for channels adjacent to incumbent, 600 MHz guard band, and duplex gap
In-band channel PSD limit (100 kHz)	Conducted power limit → conducted PSD limit 10 dBm → −7.4 dBm 14 dBm → −3.4 dBm 18 dBm → 0.6 dBm 22 dBm → 4.6 dBm 26 dBm → 8.6 dBm 30 dBm → 12.6 dBm
Adjacent channel PSD limit (100 kHz)	Conducted power limit → conducted PSD limit 10 dBm → −62.8 dBm 14 dBm → −58.8 dBm 18 dBm → −54.8 dBm 22 dBm → −50.8 dBm 26 dBm → −46.8 dBm 30 dBm → −42.8 dBm
Permissible TV channels	2–51
Permissible TV frequencies (MHz)	54–72, 76–88, 174–216, 470–698
Maximum time to redetermine channel availability	20 min

The new proposed changes to the operation parameters of fixed device and personal portable devices are provided in Tables 2.4 and 2.5, respectively.

The required separation distances are revised such that fixed devices with power level lower than 4 W are able to operate closer to television station protection contour. The new required separation distances for co-channel are from transmit power of 40 mW to 4 W, while for adjacent channel are from transmit power of 100 mW to 4 W since ERIP of 40 mW or less are not required to meet adjacent

Table 2.5 New Proposed Changes to Operation Parameters of Personal Portable Devices

Device Type	Mode II	Mode I
Transmit power limit (EIRP)	100 mW, 40 mW for channels adjacent to incumbent, 600 MHz guard band, and duplex gap	
Permissible TV channels	14–36, 38–51	
Permissible TV frequencies (MHz)	470–608, 614–698	
Maximum time to redetermine channel availability	20 min	NA

Table 2.6 Required Separation Distances from Protection Contour of Co-Channel Digital or Analog TV Transmitter

Antenna Height above Average Terrain of Unlicensed Device (m)	16 dBm, 40 mW (km)	20 dBm, 100 mW (km)	24 dBm, 250 mW (km)	28 dBm, 625 mW (km)	32 dBm, 1.6 W (km)	36 dBm, 4 W (km)
Personal/portable	1.3	1.7	NA	NA	NA	NA
Height < 3	1.3	1.7	2.1	2.7	3.3	4.0
3 ≤ height < 10	2.4	3.1	3.8	4.8	6.1	7.3
10 ≤ height < 30	4.2	5.1	6.0	7.1	8.9	11.1
30 ≤ height < 50	5.4	6.5	7.7	9.2	11.5	14.3
50 ≤ height < 75	6.6	7.9	9.4	11.1	13.9	18.0
75 ≤ height < 100	7.7	9.2	10.9	12.8	17.2	21.1
100 ≤ height < 150	9.4	11.1	13.2	16.5	21.4	25.3
150 ≤ height < 200	10.9	12.7	15.8	19.5	24.7	28.5
200 ≤ height ≤ 250	12.1	14.3	18.2	22.0	27.3	31.2

channel separation distances. The required separation distances of co-channel and adjacent channel from TV transmitter's protection contour are shown in Tables 2.6 and 2.7.

Wireless medical telemetry services are used for remote monitoring of patients' vital signs and other health parameters inside medical facilities, which operate in channel 37. There are also 10 very long baseline arrays (VLBAs) used by National Radio Astronomy Observatory to examine and study radio waves of cosmic origin, which operate in channel 37. The fixed devices are required to protect the operations of these services operating in channel 37. In the new proposed rulemaking,

Table 2.7 Required Separation Distances from Protection Contour of Adjacent Channel Digital or Analog TV Transmitter

Antenna Height above Average Terrain of Unlicensed Device (m)	20 dBm, 100 mW (km)	24 dBm, 250 mW (km)	28 dBm, 625 mW (km)	32 dBm, 1.6 W (km)	36 dBm, 4 W (km)
Personal/portable	0.1	NA	NA	NA	NA
Height < 3	0.1	0.1	0.1	0.1	0.2
3 ≤ height < 10	0.1	0.2	0.2	0.2	0.3
10 ≤ height < 30	0.2	0.3	0.3	0.4	0.5
30 ≤ height < 50	0.3	0.3	0.4	0.5	0.7
50 ≤ height < 75	0.3	0.4	0.5	0.7	0.8
75 ≤ height < 100	0.4	0.5	0.6	0.8	1.0
100 ≤ height < 150	0.5	0.6	0.8	0.9	1.2
150 ≤ height < 200	0.5	0.7	0.9	1.1	1.4
200 ≤ height ≤ 250	0.6	0.8	1.0	1.2	1.5

Table 2.8 Required Separation Distances from Co-Channel VLBA Sites

16 dBm, 40 mW (km)	20 dBm, 100 mW (km)	24 dBm, 250 mW (km)	28 dBm, 625 mW (km)	32 dBm, 1.6 W (km)	36 dBm, 4 W (km)
51	73	105	151	219	314

separation distances are being defined such that the fixed devices are operating far enough from wireless medical telemetry services to limit their interference. The required co-channel separation distances of fixed devices to VLBA sites are provided in Table 2.8. The required co-channel and adjacent channel separation distances from wireless medical telemetry services sites are shown in Tables 2.9 and 2.10, respectively.

The repurposed 600 MHz band is the frequency that will be reallocated and reassigned for Part 27 600 MHz services. However, FCC permits the continued use of WSDs in the repurposed 600 MHz band as it expects that some of the repurposed 600 MHz band spectrum may not be used by licensed wireless services in some areas for a considerable amount of time. The FCC, therefore, defined minimum separation distances to protect the licensed wireless broadband services in the repurposed 600 MHz band. The protection levels for the licensed wireless uplink and downlink are different; hence, there are co-channel and adjacent channel separation distances for both wireless uplink and downlink. The required separation distances for co-channel and adjacent channel wireless uplink are provided in Tables 2.11 and 2.12, respectively. The required separation distances for co-channel and adjacent channel wireless downlink are provided in Table 2.13 that do not differentiate the WSDs' antenna height.

Table 2.9 Required Separation Distances from Co-Channel Wireless Medical Telemetry Services

Antenna Height above Average Terrain of Unlicensed Device (m)	16 dBm, 40 mW (km)	20 dBm, 100 mW (km)	24 dBm, 250 mW (km)	28 dBm, 625 mW (km)	32 dBm, 1.6 W (km)	36 dBm, 4 W (km)
Height < 3	0.3	0.4	0.5	0.6	0.8	1.0
3 ≤ height < 10	0.6	0.7	0.9	1.1	1.4	1.7
10 ≤ height < 30	1.0	1.2	1.5	1.9	2.7	2.9
30 ≤ height < 50	1.2	1.6	2.1	2.4	3.0	3.8
50 ≤ height < 75	1.5	1.9	2.4	2.9	3.6	4.5
75 ≤ height < 100	1.7	2.2	2.7	3.3	4.2	5.3
100 ≤ height < 150	2.1	2.7	3.1	3.8	5.0	6.5
150 ≤ height < 200	2.5	3.1	3.4	4.3	5.8	7.4
200 ≤ height ≤ 250	2.8	3.5	3.7	4.7	6.3	8.0

Table 2.10 Required Separation Distances from Adjacent Channel Wireless Medical Telemetry Services

Antenna Height above Average Terrain of Unlicensed Device (m)	16 dBm, 40 mW (km)	20 dBm, 100 mW (km)	24 dBm, 250 mW (km)	28 dBm, 625 mW (km)	32 dBm, 1.6 W (km)	36 dBm, 4 W (km)
Height < 3	0.1	0.1	0.1	0.1	0.1	0.1
3 ≤ height < 10	0.1	0.1	0.1	0.1	0.1	0.1
10 ≤ height < 30	0.1	0.1	0.1	0.1	0.2	0.2
30 ≤ height < 50	0.1	0.1	0.1	0.2	0.2	0.3
50 ≤ height < 75	0.1	0.1	0.2	0.2	0.3	0.3
75 ≤ height < 100	0.1	0.2	0.2	0.2	0.3	0.4
100 ≤ height < 150	0.2	0.2	0.2	0.3	0.4	0.5
150 ≤ height < 200	0.2	0.2	0.3	0.3	0.4	0.6
200 ≤ height ≤ 250	0.2	0.2	0.3	0.4	0.5	0.6

Table 2.11 Required Separation Distances from Co-Channel Uplink Band and 600 MHz Band Base Stations

Antenna Height above Average Terrain of Unlicensed Device (m)	16 dBm, 40 mW (km)	20 dBm, 100 mW (km)	24 dBm, 250 mW (km)	28 dBm, 625 mW (km)	32 dBm, 1.6 W (km)	36 dBm, 4 W (km)
Height < 3	5	6	7	9	12	15
3 ≤ height < 10	9	11	14	17	22	27
10 ≤ height < 30	15	19	24	30	38	47
30 ≤ height < 50	20	24	31	38	49	60
50 ≤ height < 75	24	30	37	47	60	60
75 ≤ height < 100	27	34	43	54	60	60
100 ≤ height < 150	33	42	53	60	60	60
150 ≤ height < 200	39	49	60	60	60	60
200 ≤ height ≤ 250	43	54	60	60	60	60

Table 2.12 Required separation Distances from Adjacent Channel Uplink Band and 600 MHz Band Base Stations

Antenna Height above Average Terrain of Unlicensed Device (m)	16 dBm, 40 mW (m)	20 dBm, 100 mW (m)	24 dBm, 250 mW (m)	28 dBm, 625 mW (m)	32 dBm, 1.6 W (m)	36 dBm, 4 W (m)
Height < 3	112	141	177	223	282	354
3 ≤ height < 10	204	257	323	407	514	646
10 ≤ height < 30	354	445	560	704	890	1120
30 ≤ height < 50	457	575	723	909	1150	1446
50 ≤ height < 75	560	704	885	1113	1408	1770
75 ≤ height < 100	646	813	1022	1285	1626	2044
100 ≤ height < 150	792	996	1252	1574	1991	2504
150 ≤ height < 200	914	1150	1446	1818	2299	2891
200 ≤ height ≤ 250	1022	1285	1616	2033	2571	3232

Table 2.13 Required Separation Distances from Co-Channel and Adjacent Channel Downlink Band and 600 MHz Band Handsets

	Co-Channel (km)	Adjacent Channel (km)
Distance from the edge of handsets' base station	35	31

2.1.2 Canada: Industry Canada

In 2011, the regulator of Canada, Industry Canada, released a consultation paper on the policy and technical framework for the use of nonbroadcasting applications in the TV bands [9]. The objectives are to seek comments on the possibility of introducing WSDs in TV bands and possible changes to rules for licensed rural broadband systems and licensed low-power apparatus such as wireless microphones. Like many other countries, Canada is also switching from analog to digital TV broadcasting that will reduce the number of TV channels needed for TV broadcasting. This will increase the spectrum available for license–exempt applications so as to maximize the economic and social benefits from these services. Industry Canada expects the benefits to come from increased ranges of Wi-Fi devices, lower cost of wireless broadband equipment, and potential new applications. However, measures must still be introduced to reduce the impact on existing TV users from license–exempt applications.

In the consultation paper, Industry Canada is considering frequency bands 54–72 MHz, 76–88 MHz, 174–216 MHz, 470–608 MHz, and 614–698 MHz, which are similar to the United States. This is because spectrum above 698 MHz is currently being used by commercial entities and public safety agencies for mobile wireless communications. There is a unique system in Canada that operates in TV channels 21–51, which is the licensed subscriber-based remote rural broadband system (RRBS). The RRBS operates on a no-protection, no-interference basis with respect to the TV broadcast stations. The WSDs are required to protect the existing RRBS as well during their operations. Channel 37 is proposed not to be used by WSDs, similar to FCC's first proposed rulemaking rules so as to protect the radio astronomy and medical telemetry systems.

In the consultation paper, Industry Canada asked for comments such as the methods to prevent interference to the incumbent TV users, although it proposed that the use of database will be the main focused method. Therefore, Industry Canada will focus on addressing the parameters related to the protection of incumbent TV users for WSDB. The digital TV transmitters' signal levels that defined the protection contour are proposed in the consultation paper and are slightly different from those of the United States. In this stage, the separation distances were set to be the same as the second report and order in FCC, but are changed in the later decisions as FCC changes their separation distances too.

Spectrum sensing is being considered but Industry Canada has decided to wait for this technology to mature so that the probabilities to detect receive-only stations and hidden node stations are improved. The hybrid method that combined WSDB and spectrum sensing is also a possible intermediate step.

The Industry Canada proposed classifications of WSDs are similar to the FCC, except that they do not include sensing-only WSDs. The three types of WSDs that are similar to FCC are fixed WSD, personal portable mode I WSD, and mode II WSD. The technical parameters for WSDs are not established, but are expected to, in general, harmonize with the WSDs of the United States. This will result in lowering the cost of the equipment.

After receiving feedbacks on the consultation paper from relevant shareholders, Industry Canada has proposed to allow TVWS operations in Canada and the framework for TVWS operations is finalized and released in 2012 [10]. However, Industry Canada has stated in the report that it will make appropriate changes to regulations whenever necessary, especially when there are more results from TVWS operations around the world. The framework mostly harmonized with the rules by FCC in second memorandum opinion and order. Some differences are the exclusion of sensing-only device as its WSD-type category and no channel being reserved for wireless microphones. Industry Canada will determine its own technical rules for WSDs with its own established processes and consultations with stakeholders as the next step.

In 2015, Industry Canada completed and published the technical rules for WSDs after consultation with their stakeholders [11]. The specifications for WSDB were also published at the same time [12]. The operation parameters of WSDs determined by Industry Canada are summarized and provided in Table 2.14.

Table 2.14 Operating Parameters of WSDs under Regulation of Industry Canada

Devices	Fixed Devices	Mode II	Mode I
Transmit power limit (EIRP)	4 W; not allowed for channels adjacent to incumbent	100 mW; 40 mW for channels adjacent to incumbent	100 mW; 40 mW for channels adjacent to incumbent
In-band channel PSD limit (100 kHz)	12.6 dBm	2.6 dBm; −1.4 dBm for channels adjacent to incumbent	2.6 dBm; −1.4 dBm for channels adjacent to incumbent
Adjacent channel PSD limit (100 kHz)	−42.8 dBm	−56.8 dBm	−56.8 dBm
Band edge PSD limit (100 kHz to channel edge)	−42.8 dBm	−52.8 dBm	−52.8 dBm

Field strength emission limit at 1 m (dBµV/m/120 kHz)	Frequency (MHz)	Field strength
	602–607	120-5(F-602)[a]
	607–608	95
	608–614	30
	614–615	95
	615–620	120-5(620-F)[a]

(*continued*)

Table 2.14   (*Continued*)

Devices	Fixed Devices	Mode II	Mode I
Permissible TV channels	2, 5–36, 38–51	21–36, 38–51	21–36, 38–51
Permissible TV frequencies (MHz)	54–60, 76–88, 174–216, 470–608, 614–698	512–608, 614–698	512–608, 614–698
Discover available channel	WSDB access	WSDB access	From fixed device or mode II device
Maximum time to redetermine channel availability	Once a day	Once a day or move more than 100 m	Once every minute
Geolocation capability	Accuracy ± 50 m	Accuracy ± 50 m; Recheck location once every minute	NA
Separation distance from wireless microphones	1 km	400 m	400 m

[a]F is the frequency in MHz.

One of the main differences between regulations of FCC and Industry Canada is their protection to TV transmitters. First, the signal levels that define a TV transmitter's protection contour for both analog and digital TV stations are different in the UHF band as shown in Table 2.15. The propagation curve used by Industry Canada for computing the TV transmitter's signal is based on the field strength chart in Appendix 6 of Part 10 of broadcasting procedures and rules by Industry Canada [13]. Second, the separation distances are not just applied to co-channel and adjacent channels. There are some cases in which separation distances are also applied to channels called taboo channels. Taboo channels are defined as channels $N \pm 2$, $N \pm 3$, $N \pm 4$, $N \pm 7$, $N \pm 8$, $N \pm 14$, and $N \pm 15$, where N is the broadcast channel. Third, instead of having a separation distance from the nearest protection contour, there is also another separation distance from the far side of the protection

Table 2.15   Minimum TV Signal Levels within Protection Contour in Canada

Type of Station	Channel	Contour (dBμ)	Propagation Curve
Analog signal	Low VHF (2–6)	47	$F(50,90)$
	High VHF (7–13)	56	$F(50,90)$
	UHF (14–51)	$64\text{-}20\log(615/F)^{a}$	$F(50,90)$
Digital signal	Low VHF (2–6)	28	$F(50,90)$
	High VHF (7–13)	36	$F(50,90)$
	UHF (14–51)	$41\text{-}20\log(615/F)^{a}$	$F(50,90)$

[a]F is the center frequency of the TV channel in MHz.

contour. The far side of protection contour is the point on a protection contour that is 180° opposite to the direction from the TV transmitter to the WSD. Therefore, a WSD is required to meet both separation distances for a channel to be available to it. Fourth, Industry Canada defines different separation distances for different TV bands. The separation distances for fixed devices from TV protection contours are shown in Tables 2.16 and 2.17. The separation distances

Table 2.16 Required Separation Distances of Fixed Device from Nearest Protection Contour of TV Transmitter by Industry Canada

Maximum Antenna Height above Average Terrain (m)[a]	Channel Range	Required Separation from Nearest Digital TV Protection Contour (km)		Required Separation from Nearest Analog TV Protection Contour (km)	
		Co-Channel	First Adjacent Channel	Co-Channel	First Adjacent Channel and Taboo Channels
Below 3	2–6	37.0	2.0	28.1	2.2
	7–13	23.4	1.7	19.4	2.0
	14–51	14.4	1.4	11.4	1.0
4–10	2–6	37.0	2.0	28.1	2.2
	7–13	23.4	1.7	19.4	2.0
	14–51	14.4	1.4	11.4	1.0
11–30	2–6	37.0	2.0	28.1	2.2
	7–13	23.4	1.7	19.4	2.0
	14–51	14.4	1.4	11.4	1.0
31–50	2–6	47.9	2.5	36.5	2.2
	7–13	30.2	1.7	25.1	2.0
	14–51	18.7	1.4	14.7	1.0
51–75	2–6	57.5	3.1	44.8	2.6
	7–13	37.2	1.9	31.1	1.7
	14–51	23.2	1.4	18.5	1.0
76–100	2–6	63.6	3.5	51.5	2.9
	7–13	42.9	2.0	36.2	1.7
	14–51	27.0	1.4	21.4	1.0
101–150	2–6	73.1	4.2	60.7	3.4
	7–13	51.7	2.1	43.9	1.8
	14–51	33.1	1.4	26.3	1.0
151–200	2–6	80.4	4.7	67.8	3.8
	7–13	58.3	2.1	49.9	1.8
	14–51	37.1	1.4	30.3	1.0
201–250	2–6	87.0	5.1	74.0	4.0
	7–13	63.5	2.2	55.6	1.8
	14–51	40.3	1.4	33.7	1.0

[a]The maximum antenna height above average terrain is the largest value determined for eight standard radials spaced every 45° of azimuth starting from true north.

Table 2.17   Required Separation Distances of Fixed Device from Far Side of Protection Contour of TV Transmitter by Industry Canada

Maximum Antenna Height above Average Terrain (m)[a]	Channel Range	Required Separation from Far Side of Digital TV Protection Contour (km)		Required Separation from Far Side of Analog TV Protection Contour (km)	
		Co-Channel	First Adjacent Channel	Co-Channel	First Adjacent Channel and Taboo Channels
Below 3	2–6	82.1	3.5	43.0	2.3
	7–13	53.4	2.6	28.2	2.2
	14–51	35.1	1.9	21.1	1.8
4–10	2–6	82.1	3.5	43.0	2.3
	7–13	53.4	2.6	28.2	2.2
	14–51	35.1	1.9	21.1	1.8
11–30	2–6	82.1	3.5	43.0	2.3
	7–13	53.4	2.6	28.2	2.2
	14–51	35.1	1.9	21.1	1.8
31–50	2–6	90.3	4.5	55.4	3.0
	7–13	63.3	3.4	36.3	2.2
	14–51	42.2	2.5	21.1	1.8
51–75	2–6	96	5.6	64.0	3.7
	7–13	71.4	4.1	44.5	2.5
	14–51	48.3	3.0	26.1	1.8
76–100	2–6	101.3	6.4	70.1	4.2
	7–13	77.2	4.5	50.7	2.6
	14–51	53.0	3.4	30.3	1.8
101–150	2–6	110.5	7.9	79.8	5.1
	7–13	85.3	5.3	59.9	2.8
	14–51	59.8	3.9	36.2	1.8
151–200	2–6	117.9	9.0	87.2	5.8
	7–13	91.1	5.9	66.5	3.0
	14–51	64.2	4.4	40.5	1.8
201–250	2–6	124.1	10.1	93.7	6.4
	7–13	96.5	6.2	71.6	3.1
	14–51	69.0	4.7	44.0	1.8

[a]The maximum antenna height above average terrain is the largest value determined for eight standard radials spaced every 45° of azimuth starting from true north.

for fixed devices from RRBS are shown in Table 2.18. The separation distances for personal portable devices from TV protection contours are shown in Table 2.19. The separation distances for fixed devices from RRBS are shown in Table 2.20.

Table 2.18 Required Separation Distances of Fixed Device from RRBS by Industry Canada

Maximum Antenna Height above Average Terrain (m)[a]	Required Separation from RRBS Transmit Base Station (km)			Required Separation from RRBS Receive Base Station (km)		
	Co-Channel	First Adjacent Channel	Second Adjacent Channel	Co-Channel	First Adjacent Channel	Second Adjacent Channel
Height < 3	8.1	1.3	0.4	31.0	7.3	1.6
3 ≤ height < 10	8.1	1.3	–	31.0	7.3	1.6
10 ≤ height < 30	8.1	1.3	–	31.0	7.3	1.6
30 ≤ height < 50	10.2	1.6	–	38.3	9.1	1.9
50 ≤ height < 75	12.1	1.8	–	44.6	10.8	2.2
75 ≤ height < 100	14.2	2.0	–	50.1	12.7	2.5
100 ≤ height < 150	17.1	2.3	–	57.8	15.3	2.9
150 ≤ height < 200	20.3	2.7	–	64.6	18.0	3.3
200 ≤ height ≤ 250	22.8	2.9	–	69.9	20.7	2.6

[a]The maximum antenna height above average terrain is the largest value determined for eight standard radials spaced every 45° of azimuth starting from true north.

Table 2.19 Required Separation Distances of Personal Portable Device from Protection Contour of TV Transmitter by Industry Canada

Point on Protection Contour	Required Separation from Digital TV Protection Contour (km)		Required Separation from Analog TV Protection Contour (km)	
	Co-Channel	First Adjacent Channel	Co-Channel	1st Adjacent Channel and Taboo Channels
Nearest	14.4	1.1	11.4	1.0
Far side	35.1	1.9	21.1	1.8

Table 2.20 Required Separation Distances of Personal Portable Device from RRBS by Industry Canada

Maximum Antenna Height above Average Terrain (m)[a]	Required Separation from RRBS Transmit Base Station (km)		Required Separation from RRBS Receive Base Station (km)	
	Co-Channel	First Adjacent Channel	Co-Channel	First Adjacent Channel
< 250	8.2	0.7	37.8	6.5

[a]The maximum antenna height above average terrain is the largest value determined for eight standard radials spaced every 45° of azimuth starting from true north.

2.2 EUROPE

2.2.1 The United Kingdom: Ofcom

In 2005, the regulator of the United Kingdom, Ofcom, announced a Digital Dividend Review to examine the use of TV spectrum freed up by the digital switchover programme. The digital switchover programme, which is the switching of analog to digital TV broadcast, will release a large amount of spectrum for new services. Ofcom released a consultation report on the use of this free spectrum so as to maximize its value to benefit the society in 2006 [14]. In the consultation report, Ofcom indicated that they do not wish to allocate the spectrum to particular uses as they think that this proposal will reduce flexibility of the spectrum, distort competition and incentive. There is also a risk of choosing a use or user that do not provide the best benefit. Ofcom prefers to a market-led approach where Ofcom imposes as few constraints as possible on the spectrum and leave the market to determine the use of the spectrum.

After the consultation with the stakeholders, Ofcom issued a statement that proposed the use of unlicensed WSDs in the interleaved spectrum of TV bands in 2007 [15]. This is based on the reason that there are significant spectrums that are often unused at any one location at least some of the time. These spectrums can potentially be used to support a wide range of services by using new technologies such as cognitive radio that can utilize these spectrums without causing harmful interference. However, unlicensed WSDs must prove that they have the abilities to protect incumbent users before Ofcom approves their usage.

The Ofcom released consultation and statement reports in 2009 to propose and determine the technical parameters of WSDs such that they can unitize the interleaved spectrum without causing harmful interference to the incumbent users [16,17]. Three methods—spectrum sensing, WSDB, and beacon transmission—are proposed to enable WSDs to determine unused spectrum that will not cause harmful interference. The Beacon transmission method employs a network of transmitters that will broadcast beacon transmission to inform WSDs of the available TV channels within the vicinity. Of the three proposed methods, WSDB is considered the more important method in the short to medium term, while beacon technique is determined as the least appropriate and hence will not be considered further. Ofcom proposed that spectrum sensing should be allowed and proposed some key sensing parameters that WSDs should achieve, which is provided in Table 2.21. However, Ofcom is of the view that the implementation of sensing WSDs that can achieve the proposed parameters is likely many years later and hence will not determine the regulation rules of sensing WSDs for now.

As WSDB is deemed to be more important in the short to medium term, Ofcom has released further consultation and statement reports to determine the details of the parameters for WSDB [18–20]. In these reports, initial proposals and decisions on the framework and parameters for WSDs based on WSDB access are provided such as the information exchange between WSDs and WSDB, update period of WSDBs, propagation algorithms, and operation parameter algorithms.

Table 2.21 Proposed Spectrum Sensing Parameters by Ofcom

Spectrum Sensing Parameters	Value
Sensitivity with 0 dBi antenna	Digital Terrestrial Television (DTT) signal: -120 dBm (8 MHz)
	Wireless microphone signal: -126 dBm (200 kHz)
Transmit power	4–17 dBm
Out-of-band power limit	-46 dBm
Maximum time between sensing	1 s

However, Ofcom believes that practical trials and demonstrations are necessary to validate that WSDB is able to prevent harmful interference to incumbent users and will monitor the trials before allowing WSDs in the market.

The Ofcom continues their efforts in TVWS by consulting stakeholders on the technical specifications for WSDs [21]. The Ofcom classifies WSDs into two main categories: master and slave devices. Master WSDs will contact WSDB for available channels while slave WSDs will receive relevant information from their master WSDs without the need for them to contact WSDB directly. There are two types of operational parameters being defined. One of them is specific operational parameters where WSDB will derive the parameters specifically for a particular WSD based on the device's information. The other is generic operational parameters where the WSDB will derive default parameters for all slave WSDs under the control of a master WSD. Slave WSDs are able to obtain specific operational parameters from WSDB through master WSD if they are able to provide the required devices' information.

The WSDs are further divided into two types of devices, which are referred as type A and type B devices. A type A device is for fixed use and has integral, dedicated, or external antennas. A type B device is for mobile use and has an integral or dedicated antenna. The WSDs will also be classified according to their adjacent channel leakage ratios (ACLR). The ACLR will in turn determine their maximum OOB emission power limit that is provided in Equation 2.1. The Ofcom initially proposed four emission classes but later changed to five so as to be consistent with CEPT. When the type of device and the emission class are given to WSDB, the WSDB can make better decisions on the operating parameters for WSDs.

The OOB EIRP spectral density limit of a WSD proposed by Ofcom is as follows:

$$P_{\text{OOB(dBm/0.1 MHz)}} < \max\left\{P_{\text{IB(dBm/8MHz)}} - \text{ACLR(dB)}, -84\right\}, \qquad (2.1)$$

where P_{IB} is the WSD's in-band EIRP spectral density.

The Ofcom releases further consultation paper on the parameters and algorithms that will determine the available TV channels and their maximum transmit powers that can be used by WSDs such that there is a low probability of harmful interference to digital terrestrial television (DTT), Programme Making and Special Events (PMSE), and services above and below UHF TV band [22]. The proposed parameters and algorithms will be tested in pilot projects to determine their impacts to the incumbent users before any final decisions are made. The coexistence test reports on DTT and PMSE are released about a year later [23,24]. The tests for DTT coexistence include interference of WSDs to DTT receivers' at rooftop over both short and long ranges and also in laboratory conditions. The tests for PMSE coexistence include live events in theatres and outdoor broadcast, controlled trials in theaters, and also in laboratory conditions. The test results from these reports are used for the decisions of parameters and rules for the WSD operations.

In 2015, the Ofcom has given the green light to use TVWS after releasing their decisions on implementations of TVWS [25]. One of the criteria for unlicensed WSDs to operate in TVWS is that the operating parameters of WSDs must not be able to manually configure by end users. WSDs that require manual configurations may be authorized on licensed basis. Further details of TVWS regulations set by Ofcom are provided in the following paragraphs.

The framework for WSDB access proposed by the Ofcom is shown in Figure 2.4. In this framework, a master WSD would first obtain a list of available approved WSDBs from a website hosted by Ofcom. The website address is https:// tvws-databases.ofcom.org.uk. It would then query one of the approved WSDBs in the list for available TV channels and their transmission power limits based on its location and device parameters it has provided. The WSDB will provide operating parameters to the master WSD after computing them based on data provided by Ofcom. These data include PMSE data, DTT coexistence data, location agnostic data, and unscheduled adjustments data. A master WSD will need to request for generic operational parameters from the WSDB as well if it wants to form a network consisting slave WSDs. These generic operational parameters will ensure that slave WSDs within the master WSD's coverage can operate without causing interference to incumbent users. The slave devices will obtain generic operational

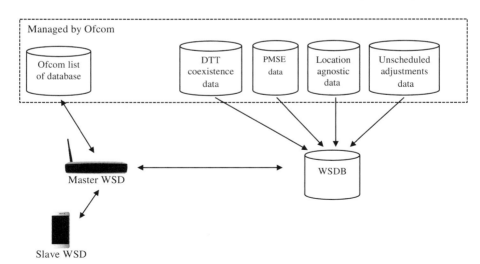

Figure 2.4 Framework of Ofcom's TV white space.

parameters from the broadcast of master WSD. If required, a slave WSD can obtain specific operating parameters after it establishes communication with its master WSD. The master WSDs are required to report the channels and power levels that both its slave WSDs and it are using to the WSDB.

The PMSE data are information on licensed PMSE that are using the TV channels other than channel 38. The DTT coexistence data are a set of data containing the maximum powers that a WSD can transmit in a $100\,m \times 100\,m$ geographic pixel in each TV channel based on the probability of harmful interference to DTT users. The location agnostic data are a set of data containing the maximum powers in each channel that does not depend on the location of the WSDs, as the location of incumbent users in such cases might not be known to WSDB. The unscheduled adjustments data are a set of transmit power limits that are introduced in an ad hoc basis at a particular location or channel.

In the TVWS framework of Ofcom, the WSDs will be provided the maximum EIRP limits in each frequency located in a geographic pixel which the WSDs are located in. The whole of the United Kingdom is divided into $100\,m \times 100\,m$ geographic pixels, which will have over 20 million of pixels. To compute the EIRP limits for the WSDs, the computing processes will be split between Ofcom and WSDBs. The Ofcom will be responsible for the computation of EIRP limits based on protection of DTT services, cross border DTT services, PMSE in channel 38, and services below $470\,MHz$ and above $790\,MHz$. It is also responsible for unscheduled adjustments of EIRP limit due to ad hoc interference issues. The WSDBs will be responsible for the computation of EIRP limits based on protection of PMSE and putting all the computed limits together taking into account the uncertainty in the location of the WSD. Details of computation processes of all the EIRP limits are provided in the annexes of [25].

An example computational process to compute the maximum in-band EIRP of a master WSD is provided as follows. The master WSD reports its horizontal location (x, y) and location uncertainty $(\Delta x, \Delta y)$. With the location information provided by the master WSD, the WSDB is able to define a set of $100\,m \times 100\,m$ geographical pixels, within which the master WSD might be located as shown in Figure 2.5. This set of pixels is defined as candidate pixels. For each candidate pixel, the WSDB will look up the power limits based on protection of DTT per available TV channel provided by Ofcom based on the emission class of the master WSD and its antenna height. The EIRP limit for each available channel will be the smallest EIRP limit value for that channel among the entire candidate pixels.

An example computational process to compute the maximum in-band EIRP of a slave WSD is provided as follows. With the location and coverage range information provided by the master WSD, the WSDB is able to define a set of $100\,m \times 100\,m$ geographical pixels, within which a slave WSD might be located as shown in Figure 2.6, where d_0 is the coverage range of master WSD and the candidate pixels are gray in color. The radius of the circle shown in Figure 2.6 is given as $d_0 + \sqrt{\Delta x^2 + \Delta y^2}$. Similarly, for each candidate pixel, the WSDB will look up the power limits based on protection of DTT per available TV channel provided by Ofcom based on the emission class of the slave WSD and its antenna height. If this

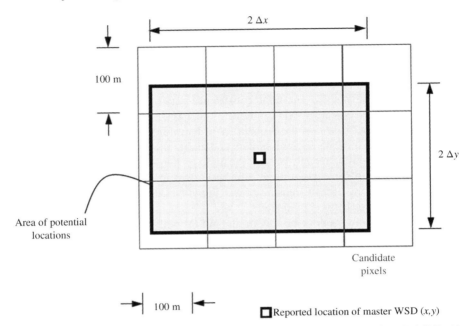

Figure 2.5 Candidate pixels where a master WSD might be located. (Reproduced from Ref. [24] with permission from Ofcom.)

information is not known to WSDB, then default values will be used. The EIRP limit for each available channel will be the smallest EIRP limit value for that channel among the entire candidate pixels.

The operating rules of master and slave devices set by Ofcom are provided in Table 2.22 and the default values of master and slave devices to be used for computation whenever the devices do not provide is given in Table 2.23.

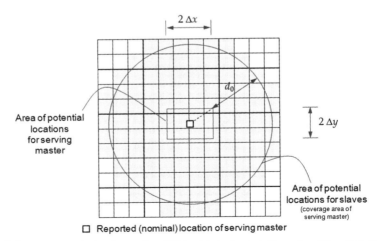

Figure 2.6 Candidate pixels where a slave WSD might be located. (Reproduced from Ref. [24] with permission from Ofcom.)

Table 2.22   Operating Rules of Master and Slave Devices Set by Ofcom

Devices	Master	Slave
Transmit power limit (EIRP)	4 W; see Table 2.25 for specific power limits in certain channels	4 W; see Table 2.25 for specific power limits in certain channels
Permissible TV channels	21–37, 39–59	21–37, 39–59
Permissible TV frequencies (MHz)	470–606, 614–782	470–606, 614–782
Discover available channel	WSDB access	From master device
Maximum time/distance to redetermine channel availability	15 min or 50 m	–
Geolocation capability	$\pm\Delta x$, $\pm\Delta y$, and $\pm\Delta z$ specified at 95% confidence level	Optional. $\pm\Delta x$, $\pm\Delta y$, and $\pm\Delta z$ specified at 95% confidence level if required
Maximum time to redetermine list of approved WSDBs	24 h	–
Maximum time to cease operation when request or lose contact	–	5 s

The WSDs will have to determine their device emission classes based on their OOB power emission levels, which in turn depends on their ACLR. The device emission classes allow the manufacturers to trade-off between device cost and channel availability. Devices with higher ACLR are more likely to be expensive to manufacture but will have better channel availability and higher transmit power limits. The five classes of emission classes approved by Ofcom are shown in Table 2.24, where class 5 will be the default class if the WSDs do not provide their class information.

Table 2.23   Default Values of WSDs Set by Ofcom

Parameter	Value
Device Emission Class	Class 5
Height of type A master WSD	20 m
Height of type B master WSD	1.5 m
Height of type A slave WSD	5 m
Height of type B slave WSD	1.5 m
Height of generic slave WSD	1.5 m
Height uncertainty	0 m, that is, no uncertainty
Technology identifier	"Generic"
Reference sensitivity of a slave WSD for computing coverage area of a master WSD	-114 dBm/100 kHz

Table 2.24   Device Emission Classes Defined by Ofcom and ETSI

Where P_{OOB} Falls within the nth Adjacent DTT Channel	ACLR (dB)				
	Class 1	Class 2	Class 3	Class 4	Class 5
$n = \pm1$	74	74	64	54	43
$n = \pm2$	79	74	74	64	53
$n \geq \pm3$	84	74	84	74	64

In order to protect the PMSE in channel 38, Ofcom has decided to disallow all WSDs to use channel 38. In addition, Ofcom also limits the WSDs maximum powers in channels 34–41 to further limit the harmful interference to the PMSE in channel 38. There are a large number of different users, with different technical characteristics, operating in the spectrum between 450 and 470 MHz, known as the UHF 2 band. There is a risk that WSDs operating in channels near them will cause harmful interference to these devices in the UHF 2 band. Therefore, Ofcom also limits the WSDs maximum powers in channels 21–24 to protect the users in the UHF 2 band. The maximum power limits of WSDs in these channels are provided in Table 2.25.

Table 2.25   Maximum Power Limits on Specific TV Channels Based on Their Device Type and Emission Class

Channels	Class 1		Class 2		Class 3		Class 4		Class 5	
	Type A (dBm)	Type B (dBm)	Type A (dBm)	Type B (dBm)	Type A (dBm)	Type B (dBm)	Type A (dBm)	Type B (dBm)	Type A (dBm)	Type B (dBm)
21	30	30	30	30	20	20	10	10	−1	−1
22	35	35	30	30	30	30	20	20	9	9
23	36	36	30	30	30	30	30	30	20	20
24	36	36	36	36	36	36	36	36	30	30
34	36	36	36	36	34	31	34	31	25	22
35	33	30	33	30	25	22	25	22	15	12
36	29	26	25	22	25	22	15	12	4	1
37	24	21	24	21	15	12	5	2	−6	−9
38				Not allowed						
39	24	21	24	21	15	12	5	2	−6	−9
40	29	26	25	22	25	22	15	12	4	1
41	33	30	33	30	25	22	25	22	15	12
42	36	36	36	36	34	31	34	31	25	22

2.2.2 Europe: CEPT

In 2011, Electronic Communications Committee (ECC) under the European Conference of Postal and Telecommunications Administrations (CEPT) released a report on technical and operational requirements for cognitive radio systems in TVWS in 2011 [27]. In this report, it concluded that sensing alone is not reliable enough to guarantee the protection of co-channel DTT receivers and PMSE, and WSDB is the most feasible method either alone or combined with sensing. The report provides details on the procedure for WSDB to calculate available channels and their maximum powers allowable by WSDs. The deployment of WSDs will be based on the master and slave architecture similar to the one proposed by Ofcom.

In 2013, the ECC released a report [28] to enhance the details of implementing the WSDB. The general principles of (1) location accuracy, (2) master–slave operations, (3) database management, (4) translation process in WSDB, and (5) combination of sensing and WSDB are provided in this report. At the same time, the ECC also released another report [29] to enhance and facilitate the development of the regulation for WSDs. The technical investigations in this report included (1) classification of WSDs, (2) collaborative spectral sensing, (3) protection parameters of DTT, (4) protection parameters of PSME, (5) protection of aeronautical radio-navigation service system, (6) impact of WSD interference on services operating at adjacent channels, and (7) potential amount of spectrum available for WSDs.

The requirements of WSD set by the ECC are as follows. Similar to the Ofcom, the ECC defines two types of WSD: the master and slave WSDs. The master WSD obtains operating parameters directly from a WSDB, while a slave WSD obtains its operating parameters from its master WSD. The WSDs are then further classified into type A and type B devices. A type A WSD has its antennas permanently mounted on a fixed outdoor installation where the location is fixed. A type B WSD has integral antenna and is not permanently mounted on a fixed outdoor installation where the location is fixed.

The requirements of master WSD set by ECC are that it must only transmit after connected to a WSDB that is approved by national regulatory authority of that country. A master WSD must be able to

- discover an approved WSDB,
- transmit its device parameters to WSDB,
- receive its operating parameters from WSDB,
- transmit its chosen channel and operating parameters back to WSDB,
- operate within the operating parameters provided by WSDB,
- manage and communicate with slave WSDs such that they are able to operate within their requirements, and
- cease transmission immediately when it moves out of its valid geographical area, expiry of given time limit or instructed by WSDB.

The device parameters that a master WSD must provide to the WSDB include

- antenna location and accuracy,
- device types (type A or B) of the master WSD and all its associated slave WSDs,
- device emission class of the master WSD and all its associated slave WSDs,
- technology identifier (e.g. LTE or IEEE 802.22) of the master WSD and all its associated slave WSDs,
- device models (of certain manufacturer) of the master WSD and all its associated slave WSDs,
- device unique identifier of the master WSD and all its associated slave WSDs, and
- device category (master or slave).

Optional device parameters that a master WSD can provide to the WSDB include

- antenna height and
- antenna angular discrimination and polarization.

The operating parameters that a master WSD should be able to receive after sending its device parameters to WSDB include

- list of available frequencies (lower and upper frequency boundaries) within one or more geographical pixel,
- maximum EIRP of the available frequencies allowed by master and its slave WSDs,
- maximum number of contiguous DTT channels and maximum number of DTT channels that master WSD and its slave WSDs can transmit, and
- time validity of the operating parameters provided by the WSDB.

Optional operating parameters that a master WSD should be able to receive after sending its device parameters to WSDB include

- sensing level for detection of protected services and
- "cease operation" message.

The channel usage parameters that a master WSD must provide to the WSDB before it starts operating after receiving operating parameters from WSDB include

- selected operating frequency (lower and upper frequency boundaries),
- maximum transmit EIRP within the selected operating frequency, and
- coverage area of master WSD.

The requirements of slave WSD set by ECC are as follows:

- Transmits its device parameters to master WSD
- Receives its operating parameters from master WSD

- Transmits its chosen operating parameters back to master WSD
- Operate within the operating parameters provided by master WSD

The device parameters that a slave WSD must provide to master WSD include

- device type,
- emission class,
- technology identifier,
- model identifier, and
- unique device identifier.

Optional device parameters that a slave WSD can provide to master WSD include

- antenna location and accuracy,
- antenna height, and
- antenna angular discrimination and polarization.

The operating parameters that a slave WSD should be able to receive after sending its device parameters to master WSD include

- list of available frequencies (lower and upper frequency boundaries) where slave WSDs are allowed to operate,
- maximum EIRP of the available frequencies allowed,
- maximum number of contiguous DTT channels and maximum number of DTT channels that slave WSDs can transmit, and
- time validity of the operating parameters.

Optional operating parameters that a slave WSD should be able to receive after sending its device parameters to master WSD include

- sensing level for detection of protected services and
- "cease operation" message.

The implementation of reliable spectrum sensing to protect the incumbent services has many challenges; however, the ECC has the view that the spectrum sensing could be combined with WSDB to improve the protection of incumbent services. The view of ECC is that by combining the two techniques, it will have advantages such as reducing the risk of interference when compared to using only one technique and allowing the protection of incumbent services that are not registered in WSDB. The ECC has defined a methodology for combining the techniques of spectrum sensing with WSDB. The flowchart of the decision making of available channels by combining the two techniques is shown in Figure 2.7.

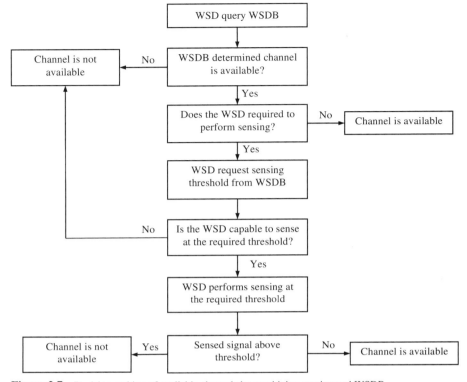

Figure 2.7 Decision making of available channels by combining sensing and WSDB.

2.3 ASIA PACIFIC

2.3.1 Singapore, IDA

Singapore is active in promoting TVWS. In 2012, Singapore White Spaces Pilot Group (SWSPG) was established with the support of Infocomm Development Authority (IDA), the national regulator for Telecommunications. The members of SWSPG include I^{2}R, Microsoft, StarHub, Neul, NICT, Power Automation, and so on. In the past 2 years, SWSPG has conducted various TVWS pilot projects and has led Singapore to be an innovation zone and test bed of advanced TVWS technologies. IDA also organizes a TVWS regulation task force under Telecommunications Standards Advisory Committee (TSAC) to set TVWS standardization in Singapore. In 2014, IDA released TVWS regulatory framework based on public feedbacks from its TVWS consultation paper [30].

IDA welcomes and intends to attract global TVWS companies coming into the Singapore market. Thus, Singapore's TVWS regulation is friendly and compatible to major TVWS standards in the world such as that of FCC, Ofcom, and CEPT. However, Singapore's TVWS regulation also has unique points that are suitable for a city country. For example, since Singapore is a small urban city, a single TV tower could cover the whole city; there are no separation distances defined for co-channel and adjacent channels. Since Singapore is a dense city, the out-of-band (OOB) emissions to adjacent channels of incumbent services are also strict. However, there are no OOB requirements for WSD-to-WSD in Singapore. To avoid the interference to the neighboring country Malaysia, IDA regulates that a WSD signal propagating at the border should be smaller than $-115\,$dBm [30].

The Singapore TVWS regulation also introduces a special type of channel called high priority channel (HPC). With HPC, WSDs could potentially have quality-of-service (QoS) differentiations even among WSDs. The detailed operation of HPC is currently under investigation with potential WSDB operators together with IDA.

The operation parameters of WSDs determined by IDA are summarized and provided in Table 2.26.

2.3.2 New Zealand, Radio Spectrum Management

The Radio Spectrum Management of New Zealand has decided to create an interim licensing scheme to allow trials of WSDs in the UHF TV bands. In 2014, it released a consultation paper for certification and licensing rules of TVWS devices [31]. A long-term regime that suits New Zealand will be investigated only after international frameworks and regulations are developed.

Under the current interim licensing scheme, WSDs have to follow either FCC Part 15, Subpart H standard [32] or the standard "ETSI EN 301 598 V1.1.1" published by the European Telecommunications Standards Institute (ETSI) [33]. In the consultation paper, it is proposed that WSDs are able to transmit a maximum EIRP of $10\,$dBW.

2.4 REGULATION COMPARISON

In this section, we compare the regulatory specifications, in particular, spectrum usage, power limit, protection level, and WSDB requirements among different countries.

2.4.1 TVWS Frequency Range

The recommended frequency ranges for TVWS regulations in different countries are listed in Table 2.27. From the table, one could see that the frequency ranges are similar for countries in the same region. This is probably due to history of TV band frequency usage. From the table, it can be observed that the TVWS frequency ranges of the United States, Canada, and Singapore are very wide spanning from VHF to UHF. While in Europe and New Zealand, the TVWS frequency ranges are only in the UHF. Large frequency range may be a challenge for antenna design and

Table 2.26 Operating Parameters of WSDs under Regulation of IDA

Devices	Fixed Devices	Mode II	Mode I
Transmit power limit (EIRP)	4 W; not allowed for channels adjacent to incumbent	100 mW	100 mW
Adjacent channel PSD limit (100 kHz)	−56.8 dBm	−56.8 dBm	−56.8 dBm
Permissible TV channels	Before analog switch off:6, 10, 11, 25, 26, 39, 41–50, 52–54, 56–58, 61, 62; after analog switch off: 5, 6, 8, 10–12, 21–28, 39–48	Before analog switch off: 6, 10, 11, 25, 26, 39, 41–50, 52–54, 56–58, 61, 62; after analog switch off: 5, 6, 8, 10–12, 21–28, 39–48	Before analog switch off: 6, 10, 11, 25, 26, 39, 41–50, 52–54, 56–58, 61, 62; after analog switch off: 5, 6, 8, 10–12, 21–28, 39–48
Permissible TV frequencies (MHz)	Before analog switch off: 181–188, 209–223, 502–518, 614–622, 630–710, 718–742, 750–774, 790–806; after analog switch off: 174–188, 195–202, 209–230, 470–534, 614–694	Before analog switch off: 181–188, 209–223, 502–518, 614–622, 630–710, 718–742, 750–774, 790–806; after analog switch off: 174–188, 195–202, 209–230, 470–534, 614–694	Before analog switch off: 181–188, 209–223, 502–518, 614–622, 630–710, 718–742, 750–774, 790–806; after analog switch off: 174–188, 195–202, 209–230, 470–534, 614–694
Discover available channel	WSDB access	WSDB access	From fixed device or mode II device
Maximum time to redetermine channel availability	6 h	6 h or move more than 100 m	Once every minute
Geolocation capability	Accuracy ± 50 m	Accuracy ± 50 m	NA

Table 2.27 TVWS Frequency Ranges for Different Regulations

FCC (MHz)	Canada (MHz)	Ofcom (MHz)	CEPT (MHz)	IDA (MHz)	New Zealand (MHz)
Fixed WSDs: 54–72, 76–88, 174–216 Fixed and portable WSDs: 470–698[a]	54–72, 76–88, 174–216, 470–608, 614–698	470–550, 614–790	470–790	181–188, 209–223, 502–518, 614–622, 630–710, 718–742, 750–774, 790–806	510–606

[a]Except areas where Part 27 600 MHz band wireless licensees operate.

size if a WSD is to cover the whole frequency range. Hence, it is important for manufacturers and users of WSDs to select the appropriate antenna for the frequency range they intend to operate.

2.4.2 Number of Channels

The comparison of the number of available channels in different regulations is shown in Table 2.28. From the table, the North American countries like the United States and Canada have more available TVWS channels than European countries. However, from Table 2.29, the channel bandwidth of each TV channel in the North American countries is 6 MHz while in Europe it is 8 MHz. Hence, the total available TVWS spectrum in North America and Europe are actually quite similar.

Singapore seems to have less TVWS spectrum compared to North America and Europe based on Tables 2.28 and 2.29. However, it is important to note that the number of TVWS channels provided in Singapore is excluding all operating TV broadcast channels, since a single TV broadcast channel will cover the whole island, thus, losing the opportunity to use this frequency as interleave spectrum. While the TVWS channels provided in North America and Europe include operating TV broadcast channels since different parts of the countries may have different number of operating TV broadcast channels. Hence, it is hard to conclude whether

Table 2.28 Number of Available Channels for Different Regulations

FCC	Ofcom	CEPT	Canada	IDA	New Zealand
50 channels; 2–51	32 channels; 21–30, 39–60	40 channels; 21–60	49 channels; 2–36, 38–51	24 channels	12 channels; 26–37

Table 2.29 The TVWS Channel Bandwidths in Different Regulations

FCC	Ofcom	CEPT	Canada	IDA	New Zealand
6 MHz	8 MHz	8 MHz	6 MHz	7 MHz (VHF) 8 MHz (UHF)	8 MHz

Singapore has more or less available TVWS spectrum compared with North America and Europe.

New Zealand has a small number of available TVWS channels as these TVWS channels are set aside for trials. It is still not known how many TVWS channels will be released if its TVWS framework is finalized.

2.4.3 Channel Bandwidth

The channel bandwidth comparison between different regulations is shown in Table 2.29. This bandwidth is consistent with the current bandwidth used by TV broadcasting.

2.4.4 Types of Devices

The device types defined in each regulation are shown in Table 2.30. In general, the types are similar, although the naming might be slightly different except for sensing-only WSDs, where currently only FCC supports this type of device. The fixed and mode II WSDs are similar to master WSDs, while mode I WSDs are similar to slave WSDs.

2.4.5 In-Channel and OOB Power Limits

In the United States, Canada, and Singapore, the in-band powers (within whole TV band) allowed by WSDs are predetermined and then based on these predetermined in-band powers, the OOB power limits (per 100 kHz) are determined. Hence, their values are fixed. However, for Europe and the United Kingdom, the maximum in-band powers of WSDs are determined by WSDB for different TV channels and

Table 2.30 The Device Types Defined in Different Regulations

FCC	Ofcom	CEPT	Canada	IDA	New Zealand
Fixed WSD, mode I WSD, mode II WSD, and sensing-only WSD	Master WSD, slave WSD	Master WSD, slave WSD	Fixed WSD, mode I WSD, and mode II WSD	Fixed WSD, mode I WSD, and mode II WSD	Fixed, base station, and mobile

Table 2.31 WSDB Requirements in Various Regulations

	FCC	Ofcom	CEPT	Singapore
Required information that WSDB has to provide to WSDs	Available TV channels	1. Start and end frequencies of available bands 2. Maximum power levels 3. Time validity of data 4. Maximum number of contiguous DTT channels 5. Maximum number of DTT channels that WSDs can transmit 6. Geographic area where operating parameters are valid 7. Maximum power limits for multichannel operation	1. Start and end frequencies of available bands 2. Maximum power levels 3. Time validity of data 4. Maximum number of contiguous DTT channels 5. Maximum number of DTT channels that WSDs can transmit	Available TV channels
WSD access frequency	20 min	2 h	–	6 h
Default time validity of data	1 h	2 h	–	6 h
Location accuracy	50 m	100 m	100 m	50 m
Reserve channels for WSDs	1	0	0	2

then the OOB limits are computed based on the provided maximum in-band powers together with the ACLR of different classes of WSDs with a minimum OOB limit set at -84 dBm/100 kHz. Hence, their values are dynamic and different across different TV channels.

2.4.6   WSDB Requirements (Table 2.31)

In general, the more frequent the WSDB accessed, the more efficient the spectrum utilization. However, there is a trade-off between complexity and efficiency. Moving forward in the future, the frequency of access might be increased.

REFERENCES

1. ITU, ITU Radiocommunication Seminar for Arab Countries, Quote from Francois Rancy, Director ITU Radiocommunication Bureau, RRS13-Arab Tunis, Tunisia, December 2013. Available at http://www.dynamicspectrumalliance.org/regulations/.
2. FCC, In the Matter of Unlicensed Operation in the TV Broadcast Bands. Additional Spectrum for Unlicensed Devices Below 900 MHz and in the 3 GHz Band, Report ET Docket No. 04–113, May 2004.
3. FCC, In the Matter of Unlicensed Operation in the TV Broadcast Bands. Additional Spectrum for Unlicensed Devices Below 900 MHz and in the 3 GHz Band, Report ET Docket No. 06–156, October 2006.
4. FCC, In the Matter of Unlicensed Operation in the TV Broadcast Bands. Additional Spectrum for Unlicensed Devices Below 900 MHz and in the 3 GHz Band, Report ET Docket No. 08–260, November 2008.
5. FCC, Radio and television broadcast rules, Title 47; CFR Part 73—Radio Broadcast Services; Subpart E—Television Broadcast Stations; §73.699: TV Engineering Charts, pp. 232–250 2012.
6. FCC, In the Matter of Unlicensed Operation in the TV Broadcast Bands. Additional Spectrum for Unlicensed Devices below 900 MHz and in the 3 GHz Band, Report ET Docket No. 10–174, September 2010.
7. FCC, In the Matter of Unlicensed Operation in the TV Broadcast Bands. Additional Spectrum for Unlicensed Devices Below 900 MHz and in the 3 GHz Band, Report ET Docket No. 12–36, April 2012.
8. FCC, In the Matter of Amendment of Part 15 of the Commission's Rules for Unlicensed Operations in the Television Bands, Repurposed 600 MHz band, 600 MHz Guard Bands and Duplex Gap, and Channel 37, and Amendment of Part 74 of the Commission's Rules for Low Power Auxiliary Stations in the Repurposed 600 MHz Band and 600 MHz Duplex Gap. Expanding the Economic and Innovation Opportunities of Spectrum Through Incentive Auctions, Report ET Docket No. 14–144, September 2014.
9. Industry Canada, Consultation on a policy and technical framework for the use of non-broadcasting applications in the television broadcasting bands below 698 MHz, SMSE-012-11, August 2011.
10. Industry Canada, Framework for the use of certain non-broadcasting applications in the television broadcasting bands below 698 MHz, SMSE-012-12, October 2012.
11. Industry Canada, White space devices (WSDs), RSS-222, Issue 1, February 2015.
12. Industry Canada, White space database specifications, DBS-01, Issue 1, February 2015.
13. Industry Canada, Broadcasting Procedures and Rules. Part 10: Application Procedures and Rules for Digital Television (DTV) Undertakings, BPR-10, Issue 1, August 2010.
14. Ofcom, Digital dividend review: This document consults on the proposed approach to the award of the digital dividend spectrum (470–862 MHz), December 2006.

15. Ofcom, Digital dividend review: a statement on our approach to awarding the digital dividend, December 2007.
16. Ofcom, Digital dividend: cognitive access—Consultation on licence–exempting cognitive devices using interleaved spectrum, February 2009.
17. Ofcom, Digital dividend: cognitive access—statement on licence–exempting cognitive devices using interleaved spectrum, July 2009.
18. Ofcom, Digital dividend: geolocation for cognitive access—a discussion on using geolocation to enable licence–exempt access to the interleaved spectrum, November 2009.
19. Ofcom, Implementing geolocation, November 2010.
20. Ofcom, Implementing geolocation: summary of consultation responses and next steps, September 2011.
21. Ofcom, TV white spaces: a consultation on white space device requirements, November 2012.
22. Ofcom, TV white spaces: approach to coexistence, September 2013.
23. Ofcom, TV White Spaces: PMSE Coexistence Tests, Technical Report, November 2014.
24. Ofcom, TV White Spaces: DTT Coexistence Tests, Technical Report, December 2014.
25. Ofcom, Implementing TV white spaces, February 2015.
26. Ofcom Implementing TV White Spaces Annexes 1 to 12, February 12, 2015.
27. ECC, Technical and Operational Requirements for the Possible Operation of Cognitive Radio Systems in the White Spaces of the Frequency Band 470–790 MHz, Report 159, January 2011.
28. ECC, Technical and Operation Requirements for the Operation of White Space Devices under Geo-Location Approach, Report 186, January 2013.
29. ECC, Complementary Report to ECC Report 159: Further Definition of Technical and Operational Requirements for the Operation of White Space Devices I in the band 470–790 MHz, Report 185, January 2013.
30. IDA, Regulatory framework for TV white space operations in the VHF/UHF bands, June 2014.
31. Radio Spectrum Management, Television white space devices certification and licensing rules, August 2014.
32. FCC, Title 47: Telecommunication; Part 15-Radio Frequency Devices; Subpart H—Television Band Devices, April 2013.
33. ETSI, White Space Devices; Wireless Access Systems Operating in the 470 MHz to 790 MHz TV Broadcast Band; Harmonized EN Covering the Essential Requirements of Article 3.2 of the R&TTE Directive, EN 301 598, v1.00.1, February 2014.

Chapter 3

Standardizations

The prospect of having a large amount of reusable spectrum combined with favorable path loss characteristics of TVWS has triggered developments of many TVWS standardizations, especially with regulations being developed at the same time. While regulations will specify the means to obtain spectrum without interfering the licensed users, standardization is developed to specify the enabling technologies of TVWS systems. Once operators obtain available TVWS spectrum under the regulation rules, they will determine which wireless standard they should adopt for their applications. For example, IEEE 802.22 is developed for long-distance communication applications, IEEE 802.11af is developed for LAN operating in TVWS spectrum, IEEE 802.15.4m is developed for WPAN focusing on M2M applications, and IEEE 802.19.1 is for coexistence between different TVWS networks. In this chapter, reviews of the IEEE 802 standards related to TVWS technology are provided.

The IEEE 802.22 is the first cognitive radio-based standard to allow sharing of geographically unused TV bands on a noninterfering basis to the incumbent users. The IEEE 802.22 working group started its activities in 2004 just after FCC's notice of proposed rulemaking in the same year. However, at the time when IEEE 802.22 was published, FCC's regulations have not been finalized. The standard is designed to support communication distance of up to 100 km.

The IEEE 802.11 family is the most popular and accepted standard in WLANs. However, WLAN systems are mostly operating in the 2.4 and 5 GHz spectrum bands, and changes in WLAN standards are required to meet the operation requirements in TV bands. The IEEE 802.11af is a modified version of existing IEEE 802.11 physical and medium access control (MAC) layers, so the WLAN systems conform to the TVWS regulations set by different countries. Therefore, IEEE 802.11af is a regulation-driven standard that needs to be flexible to accommodate the TVWS regulations of various countries.

The IEEE 802.15.4m standard is developed to enable low-rate WPAN systems to operate in TVWS. The targeted applications for IEEE 802.15.4m include M2M communications, smart grid communications, and senor networks. The main

TV White Space: The First Step Towards Better Utilization of Frequency Spectrum, First Edition.
Ser Wah Oh, Yugang Ma, Ming-Hung Tao, and Edward Peh.
© 2016 by The Institute of Electrical and Electronics Engineers, Inc. Published 2016 by John Wiley & Sons, Inc.

physical layer design of IEEE 802.15.4m is from IEEE 802.15.4g smart utility network standard.

With the various standardization activities being developed for different applications operating in TVWS, the ability of heterogeneous secondary systems to coexist is of importance. The interference between secondary systems will be made worse by the longer range that signal can propagate in TV bands. Therefore, the IEEE 802.19.1 is developed to focus on coexistence protocols and mechanisms such that the various IEEE 802 standards can share the TVWS bands without interfering with each other.

3.1 IEEE 802.19.1

3.1.1 Introduction

With the prospects of new available unlicensed spectrum and the favorable propagation characteristics, many new wireless standards are being developed to take advantage of the TVWS. However, the better propagation characteristics, which increase the transmission range, will result in more interference among TV band devices. This will make TV bands congested easily if methods to ensure coexistence among devices in TV bands are not standardized [1–3]. Regulations are defined to protect licensed devices while solutions for coexistence among unlicensed devices are left to the industry. In September 2009, the 802.19 Wireless Coexistence Technical Advisory Group started a study group on a new project, IEEE 802.19.1, to focus on coexistence methods in TV white spaces [4]. The IEEE 802.19.1 was published in 2014 [5].

The main purpose of IEEE 802.19.1 is to provide standard coexistence methods for the family of IEEE 802 wireless standards to utilize TVWS efficiently. These include dissimilar and independently operated networks and devices. An overview of the IEEE 802.19.1 will be covered in this section.

3.1.2 System Architecture

The coexistence system has three logical entities and five logical interfaces. The three logical entities are coexistence enabler, coexistence manager, and coexistence discovery and information server (CDIS). The three of the five logical interfaces are among the IEEE 802.19.1 entities, while the other two logical interfaces are for interfaces to external entities such as TVWS database and white space object (WSO). A WSO is defined as a TVWS device or network in IEEE 802.19.1. The coexistence system architecture is illustrated in Figure 3.1.

The coexistence manager is an entity that obtains all necessary information and makes coexistence decisions to provide information and management services to WSOs. The coexistence manager may also provide to a WSO information about its neighboring WSOs such as operating frequencies and transmit powers. The coexistence manager may also provide operating parameters of the WSO so that

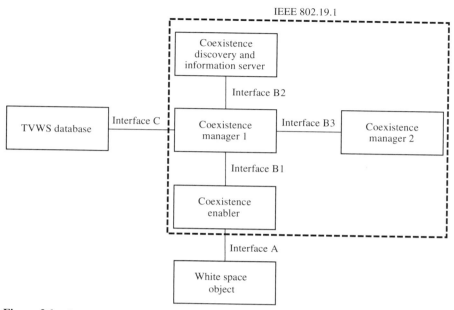

Figure 3.1   Coexistence system architecture. (Reproduced from Ref. [5] with permission from IEEE.)

the operation of the WSO is improved or prevent coexistence problems with other WSOs.

The coexistence enabler is responsible for the communications between the coexistence manager and the WSOs. It obtains information from the coexistence manager and translates them into WSO-specific information.

The CDIS provides coexistence discovery service such that coexistence managers could discover other coexistence managers in order to open interfaces between them. In order to provide the coexistence discovery service, the CDIS gathers all necessary information related to TVWS coexistence and provides them to the coexistence managers. There are two types of coexistence discovery services that CDIS may provide to the coexistence manager. The first service is called inter-CM coexistence discovery where the coexistence manager will receive coexistence set information for neighboring WSOs that are served by other coexistence managers from CDIS and then the coexistence manger will perform intra-CM coexistence discovery for WSOs under it. In the second service, the CDIS will perform both inter-CM coexistence discovery and intra-CM coexistence discovery for the coexistence manager.

The five logical interfaces are labeled as interfaces A, B1, B2, B3, and C. The interfaces B1, B2, and B3 are among the IEEE 802.19.1 entities, while interfaces A and C are between IEEE 802.19.1 entities and external systems. Interface A enables communication between coexistence enabler and WSO. The format is generic where it provides coexistence services to WSO and an interface for WSO management entity. Interface B1 is for communication between coexistence manager and coexistence enabler. Information required for coexistence and reconfiguration commands is communicated between the coexistence manager and coexistence enabler before it is forwarded to transport layer for transmission. Interface B2 is for communication between CDIS and coexistence manager where information are exchanged in order to support the coexistence discovery service that CDIS provides to coexistence manager. Interface B3 is for communication of coexistence information between coexistence managers if required. Interfaces B1, B2, and B3 shall use TCP/IP and supports secure shell protocol and transport layer security protocol. Interface C is for communication between TVWS database and coexistence manager to obtain information on available TV channels. This interface is not defined in IEEE 802.19.1 and is implementation dependent. An example of implementation of interface C is the Protocol to Access White-Space (PAWS) database [6].

For a WSO to obtain coexistence services from an IEEE 802.19.1 coexistence system, it is required to first register and subscribe coexistence services from the coexistence system. The WSO may subscribe to either information service or management service at a time. The information service allows the WSO to make autonomous coexistence decisions on its operating parameters by receiving information from the coexistence system. The management service allows the WSO to receive reconfiguration requests from the coexistence system such that the WSO can be configured based on the coexistence decision made by the coexistence system. The WSO may need to provide information or perform measurements according to the request of the coexistence system to enable the coexistence system to make coexistence decision.

3.1.3 Entities Operations

3.1.3.1 CDIS Operation

There are three different operation profiles for CDIS depending on the profile parameter provided by the coexistence manager. The operation of CDIS will consist of five states. In inactive state, the CDIS could not set up any interface connections to other IEEE 802.19.1 entities. In active state, the CDIS is able to set up interface connections to other IEEE 802.19.1 entities. In engaged state, the CDIS is ready to communicate with other IEEE 802.19.1 entities. In request sent state, the CDIS has sent out request and waiting for response. In request received state, the CDIS has received a request and is waiting for the completion of the request.

An example operation of CDIS based on profile 1 is as follows. When CDIS has done authentication with coexistence manager, it will switch to engaged state to check for subscription request from coexistence manger. The CDIS will perform coexistence manager subscription procedure and switch back to the engaged state until a registration request is received from coexistence manager. When CDIS receives the WSO registration request, it will perform the WSO registration procedure. After WSO registration, the CDIS shall perform coexistence discoveries for both intra-CM and inter-CM unless the coexistence manager subscribed only for inter-CM coexistence discovery. The CDIS will check if there are new coexistence manager requests for subscription. If there is none, the CDIS will check if current coexistence manager requires coexistence information, and if yes, it will perform the procedure to obtain the relevant information and switch back to engaged state. If a coexistence manager request for subscription change, the CDIS will perform the subscription update. When there are no more subscribed coexistence manager requests for discovery service, the CDIS will switch back to active state.

3.1.3.2 Coexistence Manager Operation

Next, the operations of coexistence manager will be briefly described. There are also three different operation profiles for coexistence manager that will depend on the neighbor coexistence manager profile parameter provided by the CDIS.

The operation of coexistence manager will consist of six states. The states are similar to the states of CDIS except that coexistence manager has one more state called waiting engagement state where the coexistence manager has requested for interface setup to another entity and is waiting for response from it.

An example operation of coexistence manager based on profile 1 is as follows:

1. After a coexistence manager authentication is done, the coexistence manager will wait for a subscription information message from the coexistence enabler so as to perform WSO subscription procedure and WSO registration procedure. It will go back to the engaged state to wait for the next message.

2. Depending on the next message, the coexistence manager will perform procedures such as sending event indication procedures, obtaining channel classification information by coexistence enabler procedure, announcing channel classification information update procedures, obtaining channel classification procedure, requesting measurement procedure, obtaining one-time measurement procedure, obtaining scheduled measurement procedure, intra-CM coexistence discovery procedure (when subscribed for inter-CM coexistence discovery only), obtaining coexistence report procedure, WSO reconfiguration/deregistration/registration/registration update/subscription update/subscription change/procedures, WSO deregistration procedure, WSO registration procedures, and coexistence manager subscription update procedure.

3. The coexistence manager will need to perform coexistence decision making when its WSO subscribed for management service. The coexistence decision making can be based on autonomous decision making where the coexistence manager makes its own decision independent of its neighbor, centralized decision making where master coexistence manager makes all decisions, including slave coexistence manager decisions, and distributed decision making where the coexistence managers make decisions through negotiation with neighboring coexistence managers.

4. When there is no coexistence enabler for the coexistence manager to serve, it will return to active state.

3.1.3.3 Coexistence Enabler Operation

The operations of coexistence enabler will be briefly described. Similarly, there are three different operation profiles for coexistence enabler and it consists of six states.

An example operation of coexistence enabler based on profile 1 is as follows:

1. After WSO authentication is done, a coexistence enabler will proceed to perform WSO subscription procedure and WSO registration procedure in sequence. It will check whether an event indication from WSO is received and will perform sending event indication from coexistence enabler to coexistence manager procedure if received.

2. When a WSO is subscribed for information service and it is requested for coexistence report, the coexistence enabler will perform obtaining coexistence report procedure. However, if no WSO is subscribed for information service but a reconfiguration request from coexistence manager is received, the coexistence enabler will perform WSO reconfiguration procedure instead.

3. At this stage, if the coexistence manager chooses to deregister, the coexistence enabler will perform WSO deregistration procedure and search for new coexistence manager to subscribe.

4. If the coexistence manager did not deregister, the coexistence enabler will perform WSO registration update procedure, obtaining channel classification information, announcing channel classification information update to coexistence enabler procedure, obtaining available channel list from WSO procedure, requesting measurement procedure, obtaining one-time measurement procedure, obtaining scheduled measurement procedure, WSO subscription update procedure, WSO subscription change procedure, and WSO registration update procedure according to the messages from WSO and coexistence manager.

3.1.4 Coexistence Mechanisms and Algorithms

The coexistence system involves two mechanisms to enable coexistence of WSOs in TVWS: coexistence management and coexistence discovery. The coexistence management is the mechanism by which the coexistence manager determines solutions for potentially interfering WSOs to share the TV channels. There are two types of services provided by the coexistence manager to the WSOs under coexistence management: information service and management service. With information service, it enables the WSOs to make their own decisions based on the information provided by the coexistence manager. In management service, the coexistence will provide the TV channels the WSOs to operate in.

The coexistence discovery mechanism is used by CDIS and coexistence manager to discover WSOs that will affect each other's performance and also enable the coexistence managers to receive information that is required for them to communicate with each other.

3.1.4.1 Coexistence Discovery Algorithms

The coexistence discovery algorithms compute whether there is harmful interference between a pair of WSOs using the same or overlapping operating frequencies, which are based on either statistical analysis of expected interference or on coverage and interference analysis.

When based on statistical analysis of expected interference, two WSOs are determined to be interfering with each other when their cumulative distribution function of potential interference measured at 90% is greater than the predetermined threshold. The information required for this algorithm includes (i) locations of WSOs, (ii) environment type (rural/urban), (iii) HAAT, (iv) supported frequencies, (v) reference bandwidth, (vi) receiver characteristics, (vii) transmitter characteristics, and (viii) antenna directivity. The algorithm will determine whether a WSO is an interference source, interference victim, or both.

When based on coverage and interference analysis, the normalized distance between two WSOs is computed based on their geographical distance and coverage regions. If the normalized distance is less than a predetermined threshold, then it is determined that the WSOs are interfering. However, the predetermined threshold can be different for different WSOs and if a WSO A is determined to interfere with another WSO B, it does not mean that WSO B is also interfering WSO A. The result of the coexistence algorithms will be sent to the coexistence managers that serve the WSOs.

3.1.4.2 Coexistence Decision Algorithms

The coexistence decision algorithms are used by coexistence manager to make decisions related to WSO reconfiguration. There are three guidelines in making the coexistence decisions, regardless of the algorithms used. The first guideline is to set different TV channels for different WSOs to avoid co-channel interference whenever it is possible. If it is not possible, similar WSOs should then be grouped together. Finally, if that is also not possible, the WSOs will share the TV channels in the time, code, or frequency domain. There are 10 types of coexistence decision algorithms defined in the standard. The algorithms are based on the following:

- Operating channel selection
- Negotiation among coexistence managers
- Resource allocation with fairness constraint
- Master coexistence manager selection
- Channel priority allocation

- Neighbor report and radio environment information
- Per-coordinate optimization
- Load balancing
- Output power control
- Co-channel sharing via WSO network geometry classification

The algorithms based on operating channel selection are based on the TV channel classification information that is in the coexistence manager. The TV channels are classified as follows:

1. Disallowed: Channels disallowed by regulation or request of incumbents
2. Allowed: Channels allowed for WSOs by regulation
3. Available: Channels that are available to use by WSOs
4. Protected: Channels that are protected due to incumbent activity
5. Restricted: Channels that have limitations on their usage by regulation
6. Operating: Channels that are being operated by WSO
7. Coexistence: Channels that are being shared by at least two WSOs as operating channels
8. Unclassified: Channels that are not classified under disallowed, available, or protected

Based on the classification information, the coexistence manager may allocate each WSO to use different TV channels if possible. The next method is to allocate WSOs that are of same type to share the same TV channels as they may have self-coexistence mechanism to mitigate co-channel interference. Finally, if a TV channel is being shared by WSOs of different types, then intersystem coexistence mechanism will need to be applied to prevent co-channel interference.

The algorithms based on negotiating among coexistence managers enable WSOs with different coexistence managers to manage their interference with each other. Etiquette mode is used such that all neighboring coexistence managers use independent operating channels. When different coexistence managers operate in the same channel, either round-robin mode can be used such that time slots are assigned sequentially to the coexistence managers or competition mode can be used such that the coexistence managers compete for time slots.

The algorithms based on resource allocation with fairness constraints can be used when fairness among the WSOs are important. The algorithms can employ Jain's fairness index and max–min algorithm to address the fairness among WSOs.

The algorithms based on master coexistence manager selection enable a coexistence manager to be the master over other coexistence managers. With a master coexistence manager, it will reduce the communication overhead among the coexistence mangers. It will also enable load balancing among the coexistence managers.

The algorithms based on channel priority allocation is intended for WSOs that have their performances degraded by interference of other WSOs but have no other available channels to use. The channel priority allocation scheme consists of two stages: coarse channel priority stage and fine priority stage. In the coarse channel priority stage, the priority allocation is based on criteria such as license type, network technology, total occupancy, and number of interfering WSOs. In the fine channel priority stage, the priority allocation will further be determined by WSOs' interference levels compared with thresholds.

The algorithm based on neighbor report and radio environment information is used to compute the operating frequencies, transmit power limits, channel shared, and transmission schedule for the WSOs. The objective of the algorithm is to enable acceptable performance for all WSOs, while minimizing the number of required reconfigurations of the WSOs. To minimize the reconfigurations, the coexistence manager will try to reconfigure the selected WSO first. Only if a solution cannot be found, the coexistence manager will try to reconfigure the WSOs that are interfering with the selected WSO. The algorithm will consider parameters such as operating frequencies, interference margins, total interference level, transmit powers, and support for scheduling when computing solutions. The solutions that do not reduce the capacity of current operating WSOs will be given priority.

The algorithm based on per-coordinate optimization uses a preset number of iterations to select the WSO operating parameters by its coexistence manager. Two objective functions can be used in this algorithm for optimization, which maximize the total throughput and minimize total interference. The coexistence manager may negotiate with neighbor coexistence managers on possible reconfiguration of their WSOs.

The algorithms based on load balancing are employed to utilize the distributed processing power of coexistence managers to maximize the number of WSOs that can be supported by the coexistence managers. The algorithm can be grouped into three methods: centralized coordination, cooperative coordination, and autonomous/distributed coordination. In centralized coordination method, a master coexistence manager will be selected and will make final decisions on operating parameters of its WSOs and all WSOs of its slave coexistence managers. In cooperative coordination method, a coexistence manager will enable its WSOs to make their own final coexistence decisions. In autonomous/distributed coordination method, the coexistence mangers will independently make their own final decisions for their WSOs.

The algorithm based on output power control is used to determine the output power of the WSOs so as to limit their aggregated interference to the protection service contour of an incumbent or a prioritized WSO. The output power of a WSO can be computed using either the flexible multiple interference margin value according to the number of active neighboring WSOs or the total aggregated in-block and out-block emission levels of the active neighboring WSOs. The selected victim reference points in the protection service contour are the points closest to each WSOs. The selected interference reference points are the closest WSOs in each network.

The algorithm based on co-channel sharing via WSO network geometry classification is used to optimize the utilization of available TV channels. There are two channel selection procedures in this algorithm. The first procedure is to assign different channels to neighboring WSOs when there are enough available channels to prevent interference among WSOs. When there are not enough channels, the algorithm will enable co-channel sharing among WSOs. For co-channel sharing, the network geometry is classified into four classes. The first class is where two different WSO network coverage areas overlap and their master WSOs are able to communicate with each other using same radio access technology that has coexistence capabilities. The second class is where two different WSO network coverage areas overlap but their master WSOs are unable to communicate even with same radio access technology. However, coexistence is still possible when slave WSOs are able to communicate with other master WSOs with same radio access technology. The third class is where different WSO network coverage areas do not overlap and all WSOs could not communicate with any WSOs in other networks even with same radio access technology. Therefore, coexistence between networks will determine by their signal-to-interference-plus-noise ratio at the edge of their expected network coverage. The fourth class is where one WSO network coverage is inside the coverage of another WSO network. The coexistence protocol will need to be checked if the radio access technology of the networks has the capabilities to coexist in this situation. Based on the network geometry classifications, the algorithm will make a final decision with three possible outcomes. The first decision is co-channel sharing by means of synchronized operation via wireless connection with similar WSOs. The second decision is no channel allocation for the selected WSO. The third decision is co-channel sharing by means of synchronized operation via backhaul connection among similar/dissimilar WSOs.

3.2 IEEE 802.22

3.2.1 Introduction

After the FCC published the proposed rulemaking to allow unlicensed wireless operation in TVWS in 2004, the IEEE 802 standards committee created the 802.22 working group on wireless regional area networks (WRANs) to utilize the TVWS [7]. The IEEE 802.22 is the first cognitive radio-based international standard to allow sharing of geographically unused TV bands on a noninterfering basis to the incumbent users such as the licensed TV broadcasters and wireless microphones [8]. This standard has been developed to operate primarily in low population density areas in order to provide broadband access to data networks for a typical coverage of 10–30 km radius from base station. However, the coverage can go up to 100 km through intelligent scheduling of the traffic to absorb additional propagation delays.

The WRAN systems will use vacant channels in the frequency range between 54 and 862 MHz. The base station is able to support up to 512 fixed units of

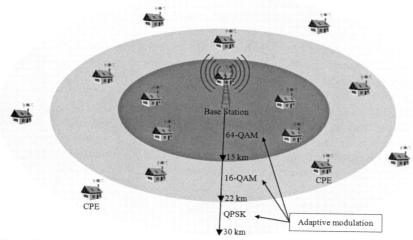

Figure 3.2 An IEEE 802.22 WRAN cell with a base station serving CPE.

customer premises equipment (CPE) with a typical VHF/UHF TV receiving antennas installation. The minimum peak throughput for downlink and uplink is
1.5 Mbps and 384 kbps, which will allow videoconferencing service [9]. An example of IEEE 802.22 WRAN cell with a base station serving CPE is shown in
Figure 3.2.

3.2.1.1 Architecture

The IEEE 802.22 architecture consists of data plane, management/control plane,
and cognitive plane. The unique characteristic of this architecture is its cognitive
components. The cognitive components must be able to determine the knowledge
of the environment, specifically which channels are occupied by the incumbents
and which channels can be used for WRAN operation. The air interface must then
be agile enough to dynamically change their channels accordingly on a real-time
basis.

The cognitive plane consists of spectrum sensing function, geolocation function, spectrum manager/spectrum sensing automation, and dedicated security sublayer 2. The spectrum sensing function is used for implementing spectrum sensing
algorithms. The geolocation function is used for providing information to determine
the location of the IEEE 802.22 devices. The spectrum manager is used to maintain
spectrum availability information, manage channel list, manage quiet periods
scheduling, and implement coexistence mechanisms. Whenever interference has
occurred, the spectrum manager is responsible for moving the operation to another
channel. To detect interference, the spectrum manager must provide sufficient quiet
period time for the CPEs to do in-band sensing on channel N and $N \pm 1$. The spectrum manager also needs to keep an updated list of backup and candidate channels

in a prioritized order. The spectrum sensing automation is a lightweight version of the spectrum manager. The spectrum sensing automation enables independent sensing procedures at initialization and start-up and also enables proper noninterfering operation of CPE during channel changes and temporary loss of communications with the base station. The security sublayer 2 is used to enhance the security for the cognitive radio-based access. Further details on the cognitive functions will be given in the next section.

The data plane consists of PHY layer, MAC layer, and the convergence sublayer. The management/control plane consists of management information base that is used for system configuration, monitoring statistics, notifications, triggers, CPE and session management, radio resource management, communication with database service, spectrum sensing and geolocation reporting, and so on.

3.2.2   Cognitive Radio Capability

As IEEE 802.22 is the first international standard that supports cognitive radio capabilities, further details on these capabilities will be provided. The cognitive radio capabilities include spectrum manager, spectrum sensing automation, access to database services, channel set management, policy, CPE registration and tracking, spectrum sensing services, and geolocation services. With these capabilities, the IEEE 802.22 devices are able to set their operating parameters according to the information from various sources such as sensing and WSDB.

3.2.2.1 Spectrum Manager Operation

The spectrum manager is the most important entity in ensuring the operations to meet regulatory requirements to protect the incumbents. It will always be present at the base station. The operation of the spectrum manager is shown in Figure 3.3.

The spectrum availability information is maintained by the spectrum manager for WRAN operation in accordance with the local regulatory policies. This

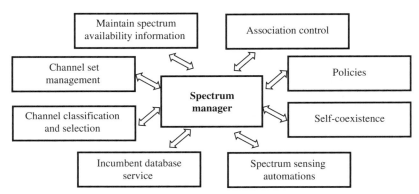

Figure 3.3   IEEE 802.22 spectrum manager operations. (Reproduced from Ref. [8] with permission from IEEE.)

information is able to be used for channel selection, channel state management, and self-coexistence mechanisms. To maintain the spectrum availability information, the spectrum manager will obtain and combine information of database services, geolocation, and spectrum sensing. The spectrum availability information will be defined during the network initialization and updated periodically throughout the WRAN operation.

In the channel classification and selection operations, other than assigning operating channels to WRAN, the spectrum manager has to maintain a list of backup channels and their corresponding priorities. The channels will be classified into one of the following categories: (i) disallowed; (ii) operating; (iii) backup; (iv) candidate; (v) protected; and (vi) unclassified. Disallowed channels are channels not allowed because of regulatory constraints. Operating channel is the current channel being used by the WRAN. Backup channels are channels that can immediately be used by WRAN in case it needs to change channel. Candidate channels are channels that have the potential to become backup channels after evaluating the incumbent status in the channels. Protected channels are channels that could not be used as the channels are occupied by the incumbents. Unclassified channels are channels that have not been sensed and the status of the incumbents is unknown.

The spectrum manager has the association control between base station and CPEs. The spectrum manager will consider the registration capabilities of CPE, the availability of channels, and EIRP limits based on data in the database before granting the association rights to the requesting CPEs to operate in a channel. The association rights must be given to CPE only when it does not cause harmful interference to the incumbents.

Policies are enforced by spectrum manager in order to protect the incumbent. The policies are related to events triggered by database services, detection of TV signals, detection of wireless microphone signals, and change of location information. For example, database services may indicate to the spectrum manager that the current operating channel is no longer available to the base station and/or some CPEs either immediately or at a specific time in the future. There will be a set of actions to be taken by the spectrum manager to ensure that the base station and CPEs will perform to protect the incumbent. These actions include (i) switching the entire WRAN cell to a new operating channel; (ii) directing CPEs to a different operating channel; and (iii) terminating operation in a given channel for CPEs or the entire cell.

A summary of the state diagram of spectrum manager is shown in Figure 3.4.

3.2.2.2 Spectrum Sensing Automation

All IEEE 802.22 devices require the spectrum sensing automation entity that interfaces with the spectrum sensing function. This entity is typically controlled by the base station except for certain conditions such as initialization. Both base station and CPE are required to perform the spectrum sensing process during initial turn-on. The base station is primarily interested in identifying an empty channel for the WRAN to operate, while a CPE is primarily interested in identifying operational WRAN channels that it can associate in this process.

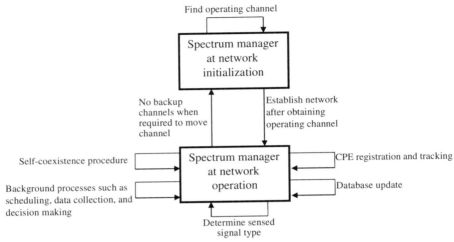

Figure 3.4   A state diagram of spectrum manager. (Reproduced from Ref. [8] with permission from IEEE.)

The spectrum sensing automation is required to perform in-band spectrum sensing and report the results to the spectrum manager. The spectrum manager will schedule quiet periods for the in-band sensing to be performed at channels N and $N \pm 1$. The sensing can be done in single time duration or in a combination of over a number of smaller time durations to reach a required sensing threshold. Other than in-band spectrum sensing, the spectrum sensing automation has to perform sensing in the backup and candidate channels periodically according to their priority levels during the idle time of CPE. The base station is responsible for allocating sufficient idle time for each CPE to perform the spectrum sensing of backup and candidate channels. The sensing results of backup channels from all spectrum sensing automations are compiled by the base station to update the backup channel list information.

When a CPE loses contact with its base station, it is the responsibility of spectrum sensing automation to make a reasonable number of attempts to reconnect with the base station. If the base station signal could not be found after a predefined duration, the CPE is required to select the next channel in the backup channel list to make attempts to reconnect with the base station. The CPE has to restart its initialization process if reassociation attempts in all the backup channels fail.

Spectrum sensing function is driven by the spectrum sensing automation. The spectrum sensing techniques include blind detectors such as energy detector, eigenvalue sensing, and multiresolution sensing. The sensing techniques for TV signal include ATSC signature sequence correlation sensing, ATSC PLL-based pilot sensing, ATSC pilot covariance sensing, ATSC cyclostationary sensing, time domain correlation-based sensing, and ATSC fast Fourier transform (FFT)-based pilot and high-order statistics sensing. Covariance-based sensing technique can be used for detecting wireless microphone. Details of these sensing techniques can be found in

Annex C of the IEEE 802.22 standard [8]. The inputs to the spectrum sensing function are radio frequency signal, country code, channel number, channel bandwidth, signal type, sensing window specification, sensing mode, and the maximum probability of false alarm. The outputs of the spectrum sensing function are signal present decision, range of confidence metric, received signal strength indicator value, range of mean and standard deviation of field strength estimate error, signal type, and sensing mode.

3.2.2.3 Geolocation and Database Services

An IEEE 802.22 device must use satellite-based geolocation techniques to determine the latitude and longitude of its transmitting antenna. The PHY and MAC of IEEE 802.22 are also able to support a two-step terrestrial-based geolocation process that is optional. The first step of the terrestrial-based geolocation is to do fine ranging between the base station and the number of its CPEs with high accuracy. The geolocation of a CPE can then be computed through triangulation based on the precise distance between the CPE and the other reference CPEs belonging to the same cell.

In a WRAN operation, the base station will interface with the database service rather than with all the individual CPEs. Therefore, the base station will be the master device that has the responsibility to protect the incumbents based on the database information, while the CPEs will be slave devices. The structure of IEEE 802.22 WRAN access to database service is shown in Figure 3.5.

The base station that is a fixed device will query the database at least once every 24 h. The base station is to provide its CPEs information such as their locations and device identification, which could be obtained from the association process, to the database. The WRAN also enables "push" technology that allows the database to send any information to the base station since the network address of the base station will be known to the database. This "push" technology function enables the WRAN to react faster to any changes in the spectrum while keeping the database traffic to a minimum.

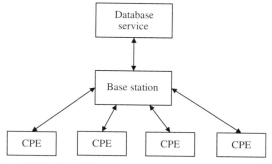

Figure 3.5   Structure of IEEE 802.22 WRAN access to database service. (Reproduced from Ref. [8] with permission from IEEE.)

3.2.3 MAC Sublayer

The IEEE 802.22 WRAN has a point-to-multipoint MAC layer and the base station manages all the activities and the associated CPEs are under its control. The downstream medium access of IEEE 802.22 where the base station transmits and CPE receives is via Time Division Multiplex, while the upstream where the CPEs transmit and base station receives is based on demands of CPEs via Demand Assigned Multiple Access with Orthogonal Frequency Division Multiple Access. The MAC supports unicast, multicast, and broadcast services. The MAC uses four different types of upstream scheduling mechanisms to control contention between CPEs within a cell and overlapping cells operating on the same channel, while at the same time meeting the latency and bandwidth requirements of each CPE. The four mechanisms are unsolicited bandwidth grants, polling, and two contention procedures.

3.2.3.1 General Superframe and Frame Structures

The IEEE 802.22 WRAN system has two operational modes and the superframe structures are different under different operational modes. In normal operational mode, there is only one WRAN in a channel and the WRAN operates on all the frames in a superframe. The second operational mode is self-coexistence mode where multiple WRANs share the same channel and each WRAN operates on one or several frames exclusively in a superframe to avoid mutual interference. A superframe is made up by 16 frames. In the normal operation mode, the first frame in a superframe consists of superframe preamble, frame preamble, superframe control header, frame header, and data payload. The rest of the 15 frames consist of frame preamble, frame header, and data payload. The superframe under normal mode is depicted in Figure 3.6. In a self-coexistence mode, the base station shall transmit its superframe preamble, frame preamble, and superframe control header at the first frame allocated to it in a superframe. If there are other frames allocated to the base

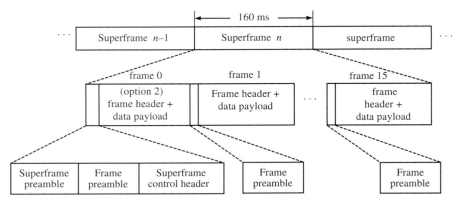

Figure 3.6   General superframe structure for normal mode. (Reproduced from Ref. [8] with permission from IEEE.)

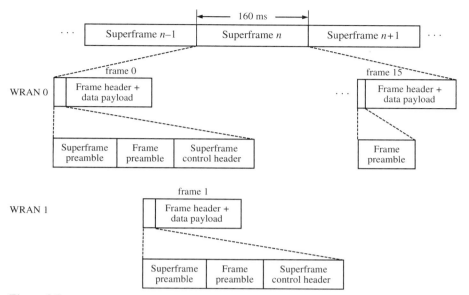

Figure 3.7   General superframe structure for self-coexistence mode. (Reproduced from Ref. [8] with permission from IEEE.)

station and its CPEs in the superframe, there is no need to retransmit the superframe preamble and superframe control header in these frames. For frames that are not allocated to the base station and its CPEs, they may monitor the channel for any transmission from neighboring WRAN cells to improve self-coexistence. The superframe under self-coexistence mode is depicted in Figure 3.7. The frame structure of the MAC is based on time division duplex where a frame is comprised of a downstream subframe and an upstream subframe. The MAC frame structure is depicted in Figure 3.8.

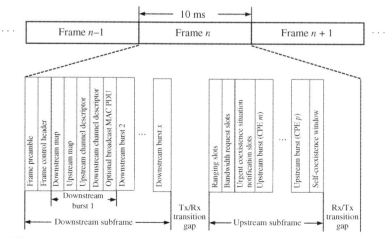

Figure 3.8   MAC frame structure. (Reproduced from Ref. [8] with permission from IEEE.)

Table 3.1 Symbol Parameters for Different Bandwidth Sizes

Bandwidths	6 MHz	7 MHz	8 MHz
Basic sampling frequency (MHz)	6.856	8	9.136
Intercarrier spacing (Hz)	3347.656	3906.25	4460.938
FFT/IFFT period, T_{FFT} (μs)	145.858	125	109.457

Source: From Ref. [8]. (Reproduced with permission of IEEE.)

The main role of the superframe control header in the superframe is to provide information in order to protect incumbents, support self-coexistence mechanisms, and support the intraframe and interframe mechanisms for management of quiet periods for sensing. The frame control header is to provide the header check sequence, length of the frame, and media access parameters. The self-coexistence window at the end of a MAC frame is to enable transmission of opportunistic coexistence beacon protocol burst.

3.2.4 Physical Layer

The PHY layer of IEEE 802.22 is based on orthogonal frequency division multiple access scheme with a FFT size of 2048. There are three channel bandwidths available that are 6, 7, and 8 MHz. The sampling frequency, intercarrier spacing, and FFT/inverse fast Fourier transform (IFFT) period for the three channel bandwidths are provided in Table 3.1. The cyclic prefix duration can be selected from one of the four values: $T_{FFT}/32$, $T_{FFT}/16$, $T_{FFT}/8$, and $T_{FFT}/4$. The total symbol duration is the FFT/IFFT period plus the cyclic prefix duration. The frequency-division multiplexing (OFDM) parameters for the three channel bandwidths are the same, having 2048 subcarriers, 1440 data subcarriers, 240 pilot subcarriers, and 368 guard subcarriers. The different PHY modes of data rate, coding rate, modulation, and the spectral efficiency of a 6 MHz channel bandwidth is provided in Table 3.2. The PHY modes 1–4 are for control signal transmission.

3.3 IEEE 802.11AF

3.3.1 Introduction

A wireless local area network (WLAN) is a wireless distribution method for linking two or more devices within a limited coverage area such as home, school, or small office. The IEEE 802.11 family standards are the most popular and accepted standard in WLANs. The IEEE 802.11 family standards have defined standards that operate in frequency bands such as 2.4, 5, and 60 GHz. However, the better propagation characteristics of TVWS band make it desirable to define a new standard of IEEE 802.11 to operate in it. Hence, the creation of IEEE 802.11af that allows WLAN operation in TVWS, which is approved in 2014 [10].

Table 3.2 Parameters of Various PHY Modes for a 6 MHZ Bandwidth with Cyclic Prefix Duration of $T_{FFT}/16$

PHY Mode	Modulation	Coding Rate	Data Rate (Mbps)	Spectral Efficiency
1	BPSK	Uncoded	–	–
2	QPSK	1/4 Repetition: 4	–	–
3	QPSK	1/3 Repetition: 3	–	–
4	QPSK	1/2 Repetition: 2	–	–
5	QPSK	1/2	4.54	0.76
6	QPSK	2/3	6.05	1.01
7	QPSK	3/4	6.81	1.13
8	QPSK	5/6	7.56	1.26
9	16-QAM	1/2	9.08	1.51
10	16-QAM	2/3	12.10	2.02
11	16-QAM	3/4	13.61	2.27
12	16-QAM	5/6	15.13	2.52
13	64-QAM	1/2	13.61	2.27
14	64-QAM	2/3	18.15	3.03
15	64-QAM	3/4	20.42	3.40
16	64-QAM	5/6	22.69	3.78

Source: From Ref. [8]. (Reproduced with permission of IEEE.)

Other than the operating frequencies, a main difference between IEEE 802.11af and the other IEEE 802.11 family standards is inclusion of WSDB mechanisms [11]. This is to conform to regulation requirements of operation in TVWS in order to protect the licensed devices. To support the WSDB mechanisms, an IEEE 802.11af network includes geolocation database dependent (GDD) enabling station, GDD dependent station, registered location secure server (RLSS), and WSDB, which is illustrated in Figure 3.9.

The GDD enabling station has the abilities to communicate with WSDB and provide its information depending on local regulation. A GDD enabling station is able to maintain operation parameters received from WSDB and use the operation parameters to control the operations of GDD dependent stations. It needs to create and transmit contact verification signal to inform GDD dependent stations of the validity of the operation parameters.

The GDD dependent station is under the control of a GDD enabling station. GDD dependent stations can query their GDD enabling station or RLSS to obtain their operation parameters. The validity of the operation parameters is verified through a contact verification signal from a GDD enabling station.

The RLSS is an entity that accesses and manages a database that organizes the storage of information by geographic location and operating parameters of one or more basic service sets. The RLSS can access to a WSDB to derive the list of available channels to requesting GDD enabling stations based on their locations. A RLSS may have the ability to map channel schedule information from TV channels to WLAN channels and provide the schedule information to GDD enabling station. A RLSS may provide GLDBs with current channel use information for all the basic

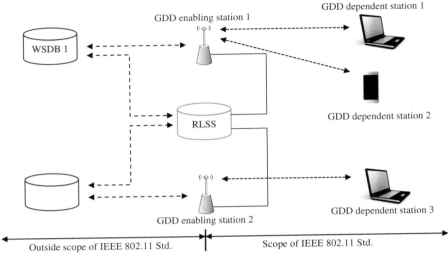

Figure 3.9   Example of IEEE 802.11af network with multiple GDD enabling stations and WSDBs. (Reproduced from Ref. [10] with permission from IEEE.)

service sets. A RLSS may also have the ability to communicate with controllers of other white space systems to coordinate emissions levels.

3.3.2   Operating Mechanisms for TVWS

To satisfy regulatory requirements in TVWS, especially in protecting the licensed devices from harmful interferences, IEEE 802.11af devices must be able to strictly operate on available TV channels at their locations provided by WSDB. The 802.11af has defined mechanisms such that the regulatory requirements can be satisfied. These mechanisms include the following:

- White space map (WSM)
- Channel available query
- Contact verification signal
- Channel schedule management
- GDD enablement
- Network channel control

The white space map is a list of available TV channels and their transmit power limits. A GDD enabling station will query a WSDB to obtain WSM information such that the GDD enabling station operates only in available TV channels indicated by WSDB. A GDD enabling station is able to generate WSMs based on the information from the GDB and update WSMs when measurements are collected. The GDD enabling station transmits WSM to GDD dependent station under its control within a GDD enablement response frame, CAQ response frame, and WSM announcement frame. A GDD dependent station operates only on channels indicated in the WSM.

The channel available query is used to obtain WSMs for an area or a geolocation for the operation of a GDD station. Both a GDD enabling station and a GDD dependent station may initiate a channel available query procedure. A GDD enabling station is required to perform channel available query procedure to update its WSM when it moves beyond the permitted distance or the validity time of its last query. A GDD enabling station can send the channel availability query frame to another GDD enabling station. In this case, the responding GDD enabling station can contact a WSDB with the location and identifying parameters sent by the requesting station. A GDD dependent station sends a channel availability query request to a GDD enabling station to obtain available TV channels. This includes the scenario that a GDD dependent station receives a contact verification signal that indicates a change of available channel list for its use. A channel available query requesting station may request available TV channels within a bounded geographic area by providing multiple geographic coordinates.

Contact verification signal is sent by a GDD enabling station for two purposes. The first purpose is to establish that GDD dependent stations are within the reception range of the GDD enabling station. The second purpose is for GDD dependent station to verify that its WSM is still valid. The contact verification signal frame has a Map ID field that indicates whether the WSM has been changed. The WSM is valid when the Map ID in contact verification signal frame is the same as the Map ID of the existing WSM. If they are different, the GDD dependent station will initiate channel available query to update its WSM.

Channel schedule management procedure is used to obtain the start and end times for each channel in the list of available TV channels. A GDD enabling station will transmit a channel schedule management request to a RLSS to query the schedule information of available TV channels. A GDD enabling station may also transmit a channel schedule management request to other GDD enabling stations that have RLSS access to obtain the TV channel schedule information. A GDD dependent station is not able to transmit a channel schedule management request.

GDD enablement is the procedure for a GDD enabling station to form a network and maintain the network under the control of a WSDB. A GDD enabling station will transmit a GDD enabling beacon signal in an available TV channel to indicate that it offers GDD enabling service. A dependent station cannot transmit any frames until it receives a GDD enabling signal where it can then transmit a GDD enablement request frame to the GDD enabling station. The GDD enabling station will send a GDD enabling response frame to indicate whether the enablement is successful.

Network channel control is the procedure to allow a responding station to manage channel usage of a requesting station. It is a two-message sequence where the first message is sent by a requesting station to request for network channel control information, which may include the requesting station's preferred channels. The responding station may provide the granted channels and their transmit power limits in the return message.

3.3.3 MAC Sublayer

The architecture of the IEEE 802.11af MAC sublayer, including the distributed coordination function, the point coordination function, the hybrid coordination function, and the mesh coordination function, follows the architecture defined in IEEE 802.11–2012, which is illustrated in Figure 3.10 [12].

The distributed coordination function is an access method known as carrier sense multiple access with collision avoidance, which will be implemented in all stations. It is used for contention services and basis for point coordination function, hybrid coordination function, and mesh coordination function.

The point coordination function is an optional access method required for contention-free services for non-QoS station under the infrastructure network configurations. The operation is similar to polling, with a point coordinator performing the role of the polling master. The point coordinator controls the frame transmissions of stations so as to eliminate contention for a limited period of time.

The hybrid coordination function an additional coordination function under the QoS facility and can only be used in QoS network configurations. This function shall be implemented in all QoS stations except for mesh stations. It allows a uniform set of frame exchange sequences to be used for QoS data transfers. The hybrid coordination function uses access methods known as enhanced distributed channel access for contention-based transfer and hybrid coordination function controlled channel access for contention-free transfer.

The hybrid coordination function controlled access is a mechanism that manages access to wireless medium that has higher medium access priority. It is required for parameterized QoS services.

The mesh coordination function is an additional coordination function under the mesh facility and can only be used in mesh basic service set. In contention-based channel access, the enhanced distributed channel access is used, while in

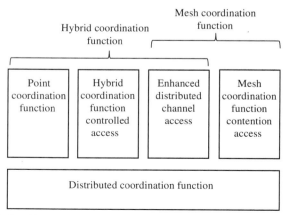

Figure 3.10 MAC architecture of IEEE 802.11. (Reproduced from Ref. [10] with permission from IEEE.)

contention-free channel access, a mechanism known as mesh coordination function controlled channel access is used.

The enhanced distributed channel access is required for prioritized QoS services and can be used by both hybrid and mesh coordination functions in contention-based channel access.

The mesh coordination function controlled channel access is an optional access method that allows mesh stations to access wireless medium with lower contention at certain times. It is required for controlled mesh service but not all mesh stations are required to use this method. This function can be used by a subset of mesh stations in a mesh basic service set.

3.3.4 Physical Layer

The IEEE 802.11af has defined a television very high-throughput (TVHT) physical specification where the basic channel units (BCUs) are 6, 7, and 8 MHz, depending on the regulatory domain. The TVHT physical specification for a BCU is based on the very high-throughput orthogonal frequency division multiplexing scheme for 40 MHz channel specified in IEEE 802.11ac [13]. The sampling clocks are changed to fit into each of the BCU bandwidths, which are provided in Table 3.3. Different bandwidth sizes are also being defined in the standard where their terminologies and definition are as provided below and illustrated in Figure 3.11.

- *TVHT_2W:* Two contiguous BCUs (12, 14, or 16 MHz)
- *TVHT_W + W:* Two noncontiguous BCUs (6 + 6, 7 + 7, or 8 + 8 MHz)
- *TVHT_4W:* Four contiguous BCUs (24, 28, or 32 MHz)
- *TVHT_2W + 2W:* Two noncontiguous frequency segments, each composed of two BCUs (12 + 12, 14 + 14, or 16 + 16 MHz)

From Table 3.3, the design is based on 144 OFDM tones in 6 and 8 MHz BCUs. The choice of 144 instead of 128 enables the signal bandwidth to reduce from 6 to 5 1/3 MHz, which will allow sharper filtering to achieve 55 dB adjacent

Table 3.3 Timing-Related Parameters for TVHT

Parameters	6 MHz	7 MHz	8 MHz
Number of data subcarriers	108	108	108
Number of pilot subcarriers	6	6	6
Total number of subcarriers	114	114	114
Highest data subcarrier index	58	58	58
Subcarrier frequency spacing	$\frac{6\,\text{MHz}}{144} = 41\frac{2}{3}\,\text{kHz}$	$\frac{7\,\text{MHz}}{168} = 41\frac{2}{3}\,\text{kHz}$	$\frac{8\,\text{MHz}}{144} = 55\frac{5}{9}\,\text{kHz}$
IFFT/FFT period	24 μs	24 μs	18 μs
Guard interval duration	6 μs	6 μs	6 μs

Source: From Ref. [3]. (Reproduced with permission of IEEE.)

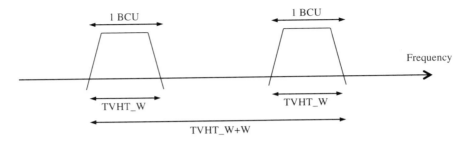

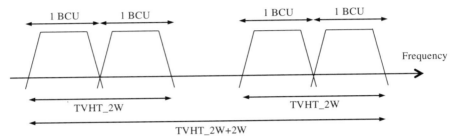

Figure 3.11   TVHT channel parameters and bandwidths. (Reproduced from Ref. [10] with permission from IEEE.)

channel leakage ratio. For 7 MHz BCU, 168 OFDM tones are chosen so as to maintain the same subcarrier frequency spacing as 6 MHz BCU.

The mandatory features that a TVHT station must support are (i) TVHT_W, (ii) single spatial stream, (iii) binary convolutional coding, and (iv) normal and short guard intervals. The optional features are (i) TVHT_2W, TVHT_W + W, TVHT_4W, and TVHT_2W + 2W, (ii) two or more spatial streams, (iii) beam-forming sounding, (iv) respond to beam-forming sounding, (v) space–time block coding, (vi) low-density parity-check code, (vii) multiuser physical layer convergence procedure protocol data unit (PPDU), and (viii) 256 quadrature amplitude modulation.

The PPDU format and the timing parameters for TVHT are provided in Table 3.4. The data rate of the TVHT will depend on the number of spatial streams

Table 3.4   PPDU Format and the Timing Parameters for TVHT

Field	6 and 7 MHz	8 MHz
Non-high-throughput short training field	60 μs	45 μs
Non-high-throughput long training field	60 μs	45 μs
Non-high-throughput signal field	30 μs	22.5 μs
TVHT signal A field (information required to interpret VHT PPDU)	60 μs	45 μs
TVHT short training field	30 μs	22.5 μs
TVHT long training field	30 μs per symbol	22.5 μs per symbol
TVHT signal B field	30 μs	22.5 μs

Source: From Ref. [3]. (Reproduced with permission of IEEE.)

Table 3.5   Data Rates of TVHT for One Spatial Stream in One BCU

Modulation	Code Rate	Coded Bits Per OFDM Symbol	Data Bits Per OFDM Symbol	Data Rate for 6 and 7 MHz (Mbps)		Data rate for 8 MHz (Mbps)	
				Guard Interval = 6 µs	Guard Interval = 3 µs	Guard Interval = 4.5 µs	Guard Interval = 22.5 µs
BPSK	1/2	108	54	1.8	2.0	2.4	2.7
QPSK	1/2	216	108	3.6	4.0	4.8	5.3
QPSK	3/4	216	162	5.4	6.0	7.2	8.0
16-QAM	1/2	432	216	7.2	8.0	9.6	10.7
16-QAM	3/4	432	324	10.8	12.0	14.4	16.0
64-QAM	2/3	648	432	14.4	16.0	19.2	21.3
64-QAM	3/4	648	486	16.2	18.0	21.6	24.0
64-QAM	5/6	648	540	18.0	20.0	24.0	26.7
256-QAM	3/4	864	648	21.6	24.0	28.8	32.0
256-QAM	5/6	864	720	24.0	26.7	32.0	35.6

Source: From Ref. [3]. (Reproduced with permission of IEEE.)

from one to four and the bandwidth size of one to four BCUs. The data rates for one spatial stream in one BCU is provided in Table 3.5, and the data rates for four spatial streams in four BCUs are provided in Table 3.6.

The transmit spectrum mask of transmission modes TVHT_W, TVHT_2W, and TVHT_4W are based on the transmit spectrum mask of IEEE 802.11ac 40 MHz mode with frequency scaling factors to fit into their respective bandwidths. For transmit spectrum mask of TVHT_2W + 2W, it is based on the transmit spectrum mask of IEEE 802.11ac 80 + 80 MHz mode with a frequency scaling factor to fit into its bandwidths. For transmit spectrum mask of TVHT_W + W, the transmit spectrum mask criterion is the same as TVHT_2W + 2W but the parameters are based on the IEEE 802.11ac 40 MHz mode parameters instead of the 80 MHz mode. The scaling factors for the transmit spectrum mask of the TVHT transmission modes are provided in Table 3.7. Examples of transmit spectral masks for TVHT_W, TVHT_W + W, and TVHT_2W + 2W for 6 MHz BCU are illustrated in Figures 3.12–3.14, respectively.

3.4 IEEE 802.15.4M

3.4.1 Introduction

Wireless personal area networks (WPANs) are designed to be used by small, inexpensive, and low power consumption wireless devices over a short distance [14]. WPANs do not require infrastructure as the communications occur in an

Table 3.6   Data Rates of TVHT for Four Spatial Streams in Four BCUS

Modulation	Code Rate	Coded Bits Per OFDM Symbol	Data Bits Per OFDM Symbol	Data Rate for 6 and 7 MHz (Mbps)		Data Rate for 8 MHz (Mbps)	
				Guard Interval = 6 μs	Guard Interval = 3 μs	Guard Interval = 4.5 μs	Guard Interval = 22.5 μs
BPSK	1/2	1728	864	28.8	32.0	38.4	42.7
QPSK	1/2	3456	1728	57.6	64.0	76.8	85.3
QPSK	3/4	3456	2592	86.4	96.0	115.2	128.0
16-QAM	1/2	6912	3456	115.2	128.0	153.6	170.7
16-QAM	3/4	6912	5184	172.8	192.0	230.4	256.0
64-QAM	2/3	10368	6912	230.4	256.0	307.2	341.3
64-QAM	3/4	10368	7776	259.2	288.0	345.6	384.0
64-QAM	5/6	10368	8640	288.0	320.0	384.0	426.7
256-QAM	3/4	13824	10368	345.6	384.0	460.8	512.0
256-QAM	5/6	13824	11520	384.0	426.7	512.0	568.9

Source: From Ref. [3]. (Reproduced with permission of IEEE.)

Table 3.7   Spectral Mask Frequency Scaling Factor for TVHT Transmission Modes

Mode	Scaling for 6 MHz	Scaling for 7 MHz	Scaling for 8 MHz
TVHT_W	6/40	7/40	8/40
TVHT_2 W	12/40	14/40	16/40
TVHT_4 W	24/40	28/40	32/40
TVHT_W + W	6/40	7/40	8/40
TVHT_2 W + 2 W	12/80	14/80	16/80

Source: From Ref. [3]. (Reproduced with permission of IEEE.)

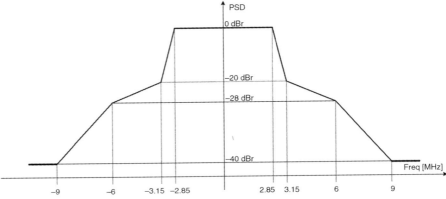

Figure 3.12   Example of transmit spectral mask for TVHT_W for 6 MHz BCU. (Reproduced from Ref. [10] with permission from IEEE.)

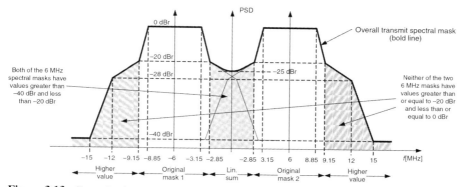

Figure 3.13 Example of transmit spectral mask for TVHT_W + W for 6 MHz BCU. (Reproduced from Ref. [10] with permission from IEEE.)

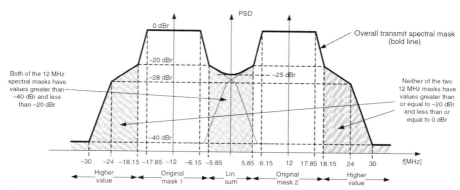

Figure 3.14 Example of transmit spectral mask for TVHT_2W + 2W for 6 MHz BCU. (Reproduced from Ref. [10] with permission from IEEE.)

ad hoc manner. There are a few standards based on the WPANs such as Bluetooth (IEEE 802.15.1), high-rate WPAN (IEEE 802.15.3), and low-rate WPAN (IEEE 802.15.4). In this section, we will focus on low-rate WPANs that are suitable in applications such as wireless sensors networks, meter reading, smart tags, home automation, and machine-to-machine communications.

As license-free spectrum becomes congested, IEEE 802.15.4 decided to explore alternative spectrum to further grow its demand; hence, IEEE 802.15.4m is introduced to the amendments of IEEE 802.15.4 for operations in TVWS. The primary purpose of IEEE 802.15.4m is to define the physical and MAC layers specifically for TVWS between 54 and 862 MHz to provide outdoor, low-data-rate wireless networks in large-scale device command and control applications [15]. The data rate ranges from 50 to 1562.5 kbps in mandatory modes and could go up to 6250 kbps in optional modes. This section surveys the IEEE 802.15.4m Network Topology, MAC and PHY layers.

3.4.1.1 *Network Topology*

The topology employed in 802.15.4m is a multichannel cluster tree topology formed by a super PAN coordinator (SPC) as the overall PAN coordinator providing synchronization services and channel allocations to other PAN coordinators in the cluster. Therefore, the SPC requires the knowledge of available TVWS channels through accessing WSDB. Each PAN cluster uses its own channel to minimize the interference among the clusters. The coverage of TVWS multichannel cluster tree PAN (TMCTP) can be expanded using a TMCTP parent–child structure. The SPC and TMCTP-parent PAN coordinator communicate with their TMCTP-child PAN coordinators during their superframe and receive beacon frames of their TMCTP-child PAN coordinators on a dedicated channel during a dedicated beacon slot (DBS) assigned. The dedicated channel is shown with an asterisk (*) in the TMCTP topology shown in Figure 3.15.

3.4.2   MAC Sublayer

3.4.2.1 *MAC Functional Description*

A coordinator on a PAN can optionally bound its channel time using a superframe structure. The superframe structure of TVWS multichannel cluster tree PAN

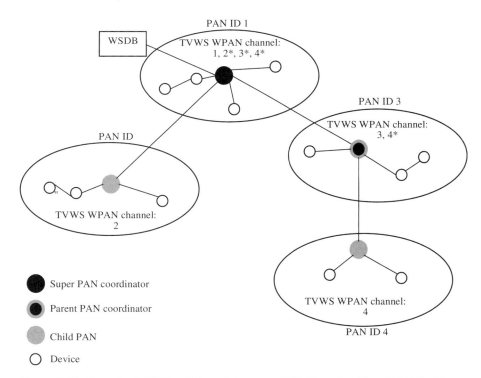

Figure 3.15   Example of TVWS multichannel cluster tree PAN. (Reproduced from Ref. [15] with permission from IEEE.)

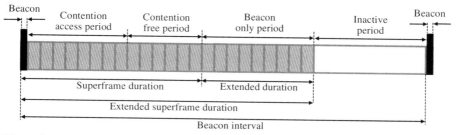

Figure 3.16 An example of TMCTP superframe structure. (Reproduced from Ref. [15] with permission from IEEE.)

(TMCTP) is composed of a beacon, a contention access period, a contention-free period, a beacon-only period, and an inactive period. The beacon is used for timing allocations and communicating management information for the PAN. The contention access period is used to transmit MAC command frames and data request command. The contention-free period contains guarantee time slots for low-latency or bandwidth-specific applications. The beacon-only period contains dedicated beacon slots used for communicating beacons between a TMCTP-parent PAN coordinator and one of its TMCTP-child PAN coordinator(s). A coordinator may enter low-power (sleep) mode during the inactive period. An example of TMCTP superframe structure is shown in Figure 3.16.

A coordinator that is not the PAN coordinator is responsible for maintaining the timing of both the incoming and outgoing superframes in a beacon-enabled PAN. Two successive frames transmitted from a device are required to be separated as the MAC sublayer needs a finite amount of time to process data. The separation time is at least an interframe spacing period, but if an acknowledgement is required between the two frames, then the separation time must be at least greater than the turnaround time.

The CSMA-CA algorithm is used for the transmission of data or MAC command frames transmitted within the contention access period, unless the frame can be quickly transmitted following the acknowledgment of a data request command. The beacon frames, acknowledgement frames, or data frames transmitted within the contention-free period are not required to use CSMA-CA algorithm. Slotted CSMA-CA algorithm is employed when periodic beacons are used; otherwise unslotted CSMA-CA algorithm is used.

3.4.2.2 TMCTP Formation Procedure

An example of the procedure to form a TMCTP formation is as follows. A SPC will obtain a list of available TVWS channels either through WSDB or from other sources. The SPC selects one of the available TVWS channels to transmit its beacon. The TMCTP-child PAN coordinator scans all the TVWS channels for the SPC's beacon and sets to SPC's channel.

The SPC transmits an enhanced beacon containing TMCTP superframe specification information elements. Upon receiving the beacon from the SPC, the

TMCTP-child PAN coordinator sends back a dedicated beacon slot (DBS) request frame to request a DBS allocation or deallocation. The SPC will generate a DBS response frame accordingly to allocate or deallocate DBS and a channel. The SPC then sends a beacon to indicate pending data for the TMCTP-child PAN coordinator, which the TMCTP-child PAN coordinator needs to reply with a data request command frame. The SPC will transmit the generated DBS response frame after receiving the data request command frame and then its own beacon frame.

The SPC switches to the channel assigned to the TMCTP-child PAN coordinator and listens for the beacon frame from the TMCTP-child PAN coordinator. The SPC will switch back to its own channel once it receives the beacon frame. If the beacon frame is not received within three beacon intervals, the SPC will switch back to its channel and retransmit the DBS response frame to the TMCTP-child PAN coordinator in the next superframe. Each TMCTP-child PAN coordinator forms an independent PAN by transmitting its beacon during its allocated DBS slot. The message sequence between the SPC and TMCTP-child PAN coordinator of this example is shown in Figure 3.17.

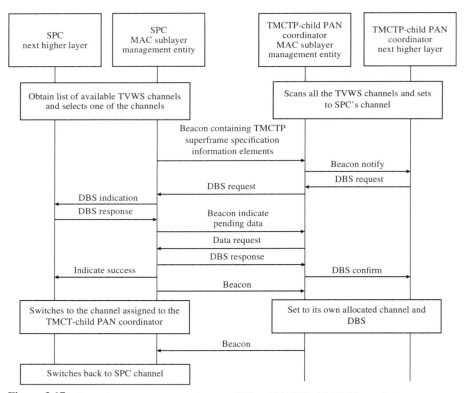

Figure 3.17   Example message sequence between SPC and TMCTP-child PAN coordinator. (Reproduced from Ref. [15] with permission from IEEE.)

Another example of the procedure to form a TMCTP between two PAN coordinators is described as follows. The TMCTP-child PAN coordinator will scan and listen for the beacon of the TMCTP-parent PAN coordinator that contains the TMCTP specification information elements. Upon successfully receiving the beacon from the TMCTP-parent PAN coordinator, the TMCTP-child PAN coordinator will send a DBS request asking for a channel and a slot. After receiving the DBS request, the TMCTP-parent PAN coordinator will either generate a DBS response frame containing the allocated/deallocated slot and channel information or it will send the DBS request frame to the SPC and receive the DBS response frame from the SPC. The TMCTP-parent PAN coordinator sends the beacon and switches to the channel allocated to TMCTP-child PAN coordinator and listens to beacon from TMCTP-child PAN coordinator before it switches back to its own channel again. The message sequence between TMCTP-parent PAN coordinator and TMCTP-child PAN coordinator of this example is shown in Figure 3.18.

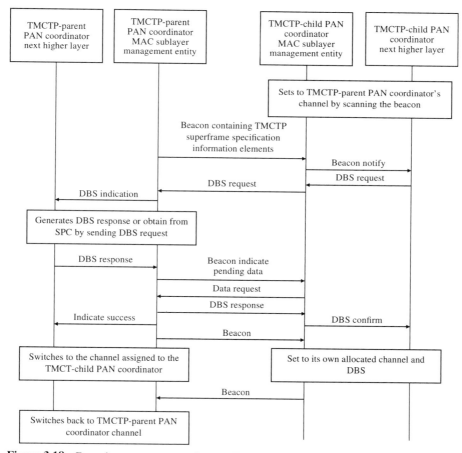

Figure 3.18 Example message sequence between TMCTP PAN coordinators. (Reproduced from Ref. [15] with permission from IEEE.)

Table 3.8   Supported Bands Field

Bit Number	Band ID	Bit Number	Band ID	Bit Number	Band ID
1	TVWS Band, USA	7	450–470 MHz	13	917–923.5 MHz
2	TVWS Band, UK	8	470–510 MHz	14	928–960 MHz
3	TVWS Band, Japan	9	779–787 MHz	15	920–928 MHz
4	TVWS Band, Canada	10	863–870 MHz	16	950–958 MHz
5	TVWS Band, Korea	11	901–902 MHz	17	2400–2483.5 MHz
6	TVWS Band, EU	12	902–928 MHz	18–23	Reserved

Source: From Ref. [15]. (Reproduced with permission of IEEE.)

An IEEE 802.15.4m device must indicate the bands that it supports through the supported bands field. This is required as the TVWS bands in different countries vary and a device that supports the TVWS bands in a country may not necessary support the TVWS bands in another country. This information is also useful in maintaining communications of the device when a band becomes unavailable. The supported bands field is shown in Table 3.8, where a value of 1 indicates that the band is supported and 0 indicates the band is not supported.

3.4.3   Physical Layer

There are three types of physical layer modes defined in 802.15.4m to support low-data-rate wireless outdoor applications using TVWS under multiple regulatory domains. These physical layer modes are designed to operate at multiple data rates in frequency bands ranging from 54 to 862 MHz. The physical layer modes are listed below.

1. Frequency Shift Keying (TVWS-FSK) PHY
2. Orthogonal Frequency Division Multiplexing (TVWS-OFDM) PHY
3. Narrow Band Orthogonal Frequency Division Multiplexing (TVWS-NB-OFDM) PHY

3.4.3.1   TVWS-FSK PHY

The PPDU structure of the TVWS-FSK PHY is shown in Figure 3.19.

	Octets	
	2	Variable
Synchronization header	PHY header	PHY payload

Figure 3.19   Format of TVWS-FSK PPDU. (Reproduced from Ref. [15] with permission from IEEE.)

Table 3.9 TVWS-FSK Modulation and Parameters

Parameter	Mode 1		Mode 2		Mode 3		Mode 4	Mode 5
Data rate (kbps)	50		100		200		300	400
Modulation level	Two-level		Two-level		Two-level		Two-level	Four-level
Modulation index- h	0.5	1.0	0.5	1.0	0.5	1.0	0.5	0.33
Channel spacing (kHz)	100	200	200	400	400	600	600	600

Source: From Ref. [15]. (Reproduced with permission of IEEE.)

Table 3.10 TVWS-FSK Symbol Encoding

Two-level		Four-level	
Symbol (Binary)	Frequency Deviation	Symbol (Binary)	Frequency Deviation
0	$-f_{dev}$	01	$-f_{dev}$
1	$+f_{dev}$	00	$-f_{dev}/3$
		10	$+f_{dev}/3$
		11	$+f_{dev}$

Source: From Ref. [15]. (Reproduced with permission of IEEE.)

The synchronization header consists of preamble field and start-of-frame delimiter (SFD) information. The preamble field consists of multiples of 8-bit sequence "01010101" and the SFD is either a 16- or 24-bit predetermined sequence. The SFD values are different for devices that support and do not support FEC. The physical header consists of ranging field, parity check field, frame check sequence-type field, data whitening field, and frame length field.

The modulation for the TVWS-FSK PHY is a two- or four-level filtered FSK where filtering is needed to meet the regulatory requirements of the frequency bands. The modulation and channel parameters are shown in Table 3.9.

The frequency deviation, f_{dev} for two-level filtered FSK is equal to (symbol rate × modulation index)/2, while for four-level filtered FSK it is equal to (3 × symbol rate × modulation index)/2. The bit-to-symbol mapping is shown in Table 3.10.

Forward error correction is optional in the TVWS-FSK PHY. There are three FEC codes defined in the standard [16]. The first FEC code is a rate 1/2 convolutional coding with constraint length $K = 7$. The second and third FEC codes are recursive and systematic codes and nonrecursive and nonsystematic codes, respectively. All the FEC codes shall be employed in conjunction with their respective interleaving scheme.

3.4.3.2 *TVWS-OFDM PHY*

The PPDU structure of the TVWS-OFDM PHY is shown in Figure 3.20.

The synchronization header consists of short training field (STF) and long training field (LTF). The STF varies from 1 to 4 OFDM symbols, while the LTF uses 2 OFDM symbols. The physical header consists of ranging field, rate field,

Number of TVWS-OFDM symbols				
Variable (3-6)	1	Variable	6 bits	Variable
Synchronization header	PHY header	PHY payload	Tail	Pad

Figure 3.20 Format of TVWS-OFDM PPDU. (Reproduced from Ref. [15] with permission from IEEE.)

frame length field, scrambler field, header check sequence field, and tail bit field. The tail and pad bits are appended for the purpose of FEC encoding.

The system parameters for mandatory and optional modes of TVWS-OFDM PHY are shown in Table 3.11. The DC subcarrier is null and the subcarriers for pilot and data tones are numbered as −54 to 54.

The FEC code employed in TVWS-OFDM PHY is a convolutional encoder of coding rate $R = 1/2$ with constraint length $K = 7$. The coded bits will go through two permutations during the interleaving process.

3.4.3.3 TVWS-NB-OFDM PHY

The PPDU structure of the TVWS-NB-OFDM PHY is shown in Figure 3.21.

The fields that are in synchronization header and physical header of TVWS-NB-OFDM are similar to TVWS-OFDM. However, the physical header of TVWS-NB-OFDM contains an additional channel aggregation field that is used for channel aggregation.

Table 3.11 System Parameters for TVWS-OFDM

Parameter	Mandatory Mode	Optional Mode
Nominal bandwidth (kHz)	1064.5	4258
Subcarrier spacing (kHz)	1250/128	$4 \times 1250/128$
DFT size	128	128
Active subcarriers	108	108
Pilot subcarriers	8	8
Data subcarriers	100	100
Data rate—BPSK (kbps)	390.625	1562.5
Date rate—QPSK (kbps)	781.250	3125
Data rate—16-QAM (kbps)	1562.5	6250

Source: From Ref. [15]. (Reproduced with permission of IEEE.)

Number of TVWS-NB-OFDM symbols		
2	1	Variable
Synchronization header	PHY header	PHY payload

Figure 3.21 Format of TVWS-NB-OFDM PPDU. (Reproduced from Ref. [15] with permission from IEEE.)

Table 3.12 System Parameters for TVWS-NB-OFDM

Parameters	Value
Nominal bandwidth (kHz)	380.95
Subcarrier spacing (kHz)	$125/126 = 0.99206$
Total number of subcarriers	384
Number of pilot subcarriers	32
Number of data subcarriers	352
Effective symbol duration (μs)	1008
Cyclic prefix interval duration (μs)	Mandatory: 1/32 (31.5 μs)
	Optional: 1/16 (63.0 μs), 1/8 (126 μs)
Data rate—BPSK (kbps)	CC code rate = 1/2:156
	CC code rate = 3/4:234
Data rate—QPSK (kbps)	CC code rate = 1/2:312
	CC code rate = 3/4:468
Data rate—16-QAM (kbps)	CC code rate = 1/2: 624
	CC code rate = 3/4: 936
Data rate—64-QAM (kbps)	CC code rate = 1/2:936
	CC code rate = 3/4:1404
	CC code rate = 7/8:1638

Source: From Ref. [15]. (Reproduced with permission of IEEE.)

The system parameters for TVWS-NB-OFDM PHY are shown in Table 3.12, where CC stands for convolutional coding.

The FEC scheme for TVWS-NB-OFDM shall consist of Reed Solomon encoding as the outer encoder and a recursive convolutional encoder of coding rate equal to 1/2, 3/4, or 7/8 as the inner encoder. The convolutional encoder encodes the Reed Solomon encoded bits together with the 6 tail bits and pad bits. All encoded bits shall be interleaved by a block interleaver defined by a two-step permutation.

TVWS-NB-OFDM supports channel aggregation to increase the utilization of the TV channels that are mostly 6 or 8 MHz. Hence, the maximum number of aggregated channels depends on the channel bandwidth. The channel aggregation parameters are shown in Table 3.13.

Table 3.13 Channel Aggregation Parameters for TVWS-NB-OFDM

Maximal Bandwidth on Channel Aggregation Use	6 MHz	8 MHz
Maximal number of subchannels available for aggregation	11	16
Channel spacing (kHz)	$(125/126 \text{ kHz} \times 404) = 400.79365$	
Guard band for each side of channel (kHz)	795.63495	793.6508

Source: From Ref. [15]. (Reproduced with permission of IEEE.)

3.5   IETF PROTOCOL TO ACCESS WHITE SPACES

PAWS is one of the interface protocols between WSD and databases developed by IETF [6]. According to the studies on multiple cognitive radio mechanisms, a simple and reliable mechanism is to use a geospatial database that contains the spatial and temporal profiles of all primary licensees' spectrum usage and requires secondary users to query the database for available spectrum that they can use at their location. Such databases can be accessible and queried by secondary users on the Internet. A high-level illustration of the databases architecture is shown in Figure 3.22. It provides a mechanism by which secondary users share the spectrum with the primary users. It is attractive in many bands and in many countries.

PAWS covers the following functions [6,17]:

1. *Database discovery:* This is a required functional component for the master device that is responsible for accessing databases for available white spaces for itself and its slave devices.

2. *Initialization:* This is a required functional component for the databases. It allows the master devices to determine necessary information yet to be preconfigured.

3. *Device registration:* This is an optional functional component for the database. It may be implemented separately or as a part of the available spectrum query functional component. It is used by the master device when the database requires due to, for example, regulatory reasons.

4. *Available spectrum query:* This is a required functional component for the master device and the database.

5. *Spectrum use notification:* This is an optional functional component for the master device and the database. If it is required, the database informs the master device to notify its spectrum use via its response to the available spectrum query.

6. *Device validation:* This is a separate functional component and optional for the master device and database. When it is implemented by the database, it allows the master device to validate slave devices without having to use the full available spectrum query.

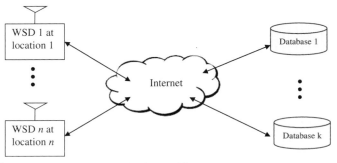

Figure 3.22   Illustration of white space database architecture.

A typical sequence of PAWS operations is as follows [6]:

1. The master device obtains the Uniform Resource Identifier (URI) for a database appropriate for its location, to which it sends subsequent PAWS messages.

2. The master device establishes an Hypertext Transfer Protocol over Transport Layer Security (HTTPS) session with the database.

3. The master device optionally sends an initialization message to the database to exchange capabilities.

4. If the database receives an initialization message, it responds with an initialization–response message in the body of the HTTP response.

5. The database may require the master device to be registered before providing service.

6. The master device sends an available spectrum request message to the database. The message may be sent on behalf of a slave device that made a request to the master device.

7. If the master device makes a request on behalf of a slave device, the master device may verify with the database that the slave device is permitted to operate.

8. The database responds with an available spectrum response message in the body of the HTTP response.

9. The master device can send a spectrum usage notification message to the database. The notification is purely informational; it notifies the database what spectrum the master device intends to use and is not a request to the database to get permission to use that spectrum. Some databases may require spectrum usage notification.

10. When the database receives a spectrum usage notification message, it responds to the master device a spectrum usage acknowledgment message. Because this notification is purely informational, the master device need not send back to the database any response.

Note that different regulatory domains could have particular requirements, for example, master devices registration to the database, slave device verification, and spectrum usage notifications.

REFERENCES

1. C. Ghosh, S. Roy, D. Cavalcanti, Coexistence challenges for heterogeneous cognitive wireless networks in TV white spaces. *IEEE Wirel. Commun.*, 18 (4), 22–31 (2011).
2. T. Baykas, S. Filin, H. Harada, Verification of TVWS coexistence system based on P802.19.1 draft standard, in Proceedings of the IEEE 77th Vehicular Technology Conference (VTC Spring), June 2013, pp. 1–5.
3. T. Baykas, J. Wang, M. A. Rahman, H. N. Tran, C. Song, S. Filin, Y. Alemseged, C. Sun, G. P. Villardi, C.-S. Sum, Z. Lan, H. Harada, Overview of TV white spaces: current regulations, standards

and coexistence between secondary users, in Proceedings of the IEEE 21st International Symposium on Personal, Indoor and Mobile Radio Communications Workshops (PIMRC Workshops), September 2010, pp. 38–42.

4. C.-S. Sum, G. P. Villardi, M. A. Rahman, T. Baykas, H. N. Tran, Z. Lan, C. Sun, Y. Alemseged, J. Wang, C. Song, C.-W. Pyo, S. Filin, H. Harada, Cognitive communication in TV white spaces: an overview of regulations, standards, and technology. *IEEE Commun. Mag.*, 51 (7), 138–145 (2013).

5. IEEE Std 802.19.1–2014, IEEE Standard for Information Technology: Telecommunications and Information Exchange Between Systems—Local and Metropolitan Area Networks—Specific Requirements—Part 19: TV White Space Coexistence Methods, pp. 1–326, June 2014.

6. E. V. Chen, S. Das, L. Zhu, J. Malyar, P. McCann, Protocol to Access White-Space (PAWS) Databases draft-ietf-paws-protocol-20, IETF, Proposed Standard, November 2014. Available at https://datatracker.ietf.org/doc/draft-ietf-paws-protocol/

7. M. Sherman, A. N. Mody, R. Martinez, C. Rodriguez, R. Reddy, IEEE standards supporting cognitive radio and networks, dynamic spectrum access, and coexistence, *IEEE Commun. Mag.*, 46 (7), 72–79, (2008).

8. IEEE Std 802.22–2011, IEEE Standard for Information Technology: Local and Metropolitan Area Networks—Specific Requirements—Part 22: Cognitive Wireless RAN Medium Access Control (MAC) and Physical Layer (PHY) Specifications: Policies and Procedures for Operation in the TV Band, July 2011, pp. 1–680.

9. C. R. Stevenson, G. Chouinard, Z. Lei, W. Hu, S. J. Shellhammer, W. Caldwell, IEEE 802.22: the first cognitive radio wireless regional area network standard. *IEEE Commun. Mag.*, 47 (1), 130–138 (2009).

10. IEEE Std 802.11af-2013, IEEE Standard for Information technology: Telecommunications and Information Exchange Between Systems—Local and Metropolitan Area Networks—Specific Requirements—Part 11: Wireless LAN Medium Access Control (MAC) and Physical Layer (PHY) Specifications Amendment 5: Television White Spaces (TVWS) Operation, February 2014, pp. 1–198.

11. A. B. Flores, R. E. Guerra, E. W. Knightly, P. Ecclesine, S. Pandey, IEEE 802.11af: a standard for TV white space spectrum sharing. *IEEE Commun. Mag.*, 51 (10), 92–100 (2013).

12. IEEE Std 802.11–2012 (Revision of IEEE Std 802.11–2007), IEEE Standard for Information technology: Telecommunications and Information Exchange Between Systems Local and Metropolitan Area Networks—Specific Requirements Part 11: Wireless LAN Medium Access Control (MAC) and Physical Layer (PHY) Specifications, March 2012, pp. 1–2793.

13. IEEE Std 802.11ac-2013, IEEE Standard for Information technology: Telecommunications and Information Exchange Between Systems—Local and Metropolitan Area Networks—Specific Requirements—Part 11: Wireless LAN Medium Access Control (MAC) and Physical Layer (PHY) Specifications—Amendment 4: Enhancements for Very High Throughput for Operation in Bands Below 6 GHz, December 2013, pp. 1–425.

14. IEEE Std 802.15.4–2011, IEEE Standard for Local and Metropolitan Area Networks—Part 15.4: Low-Rate Wireless Personal Area Networks (LR-WPANs), September 2011, pp. 1–314.

15. IEEE Std 802.15.4m-2014, IEEE Standard for Local and Metropolitan Area Networks—Part 15.4: Low-Rate Wireless Personal Area Networks (LR-WPANs)—Amendment 6: TV White Space Between 54 MHz and 862 MHz Physical Layer, April 2014, pp. 1–118.

16. IEEE Std 802.15.4k-2013, IEEE Standard for Local and Metropolitan Area Networks—Part 15.4: Low-Rate Wireless Personal Area Networks (LR-WPANs)—Amendment 5: Physical Layer Specifications for Low Energy, Critical Infrastructure Monitoring Networks, June 2013, pp. 1–149.

17. RFC 6953: Protocol to Access White-Space (PAWS) Databases—Use Cases and Requirements, Internet Engineering Task Force (IETF).

Chapter 4

TVWS Technology

TVWS is the first frequency band for real CR operation. Since there are existing PUs such as TV broadcasting and wireless microphones allocated in the same band, white space devices (WSDs) as license-exempt users have to follow special requirements such as central database access for PU protection. Accordingly, a few key technologies and their variants have been explored to meet these requirements. In this chapter, we focus our discussions on technologies that differentiate TVWS from a conventional wireless communication system. The readers could refer to standard wireless communication textbook for technologies of a typical wireless communication system, which will not be repeated in this book.

In the physical layer, TVWS covers a large frequency range. A UHF WSD may operate in frequencies from 470 to 806 MHz. If VHF TVWS is included, the frequency range could go as low as 54 MHz. This requires the TVWS antenna to have huge bandwidth. On the other hand, WSD devices are usually required to be compact. Designing wideband antenna with small size while keeping acceptable gain is a big challenge. We will address this issue with a suitable solution. To effectively protect PUs, WSDs have to be aware of PU activities in terms of the spectrum occupancies. Spectrum identification is one of the core component functions of secondary users (SUs) working in TVWS. We will introduce a few spectrum identification approaches such as spectrum sensing, WSDB, and beaconing. To avoid interfering the PU having weak received signal in the adjacent channels, WSDs have to comply with a strict out-of-band (OOB) emission limit, which is more stringent than most of the existing radio technologies such as Wi-Fi and LTE. We will discuss this challenge and the solutions. Meanwhile, since the available TVWS frequencies are usually noncontinuous fragments, where the fragments are TV channels, we will discuss channel aggregation for high-throughput transmission. According to the regulators, the fixed and mode II portable WSDs working as access points (APs) are required to look up the WSDB for channel availability with

TV White Space: The First Step Towards Better Utilization of Frequency Spectrum, First Edition.
Ser Wah Oh, Yugang Ma, Ming-Hung Tao, and Edward Peh.
© 2016 by The Institute of Electrical and Electronics Engineers, Inc. Published 2016 by John Wiley & Sons, Inc.

Table 4.1   Summary of Technical Contents in This Chapter

Application layer	REM through WSD networks
	Enhanced WSDB (area-based, embedded broadcast, WSDB-Q)
Network layer	Robust TVWS networks through advanced routing
Medium access layer	Dynamic spectrum assignment
	SU networks coexistence based on IPM
Physical layer	Positioning
	Out-of-band leakage control
	Channel aggregation
	Spectrum identification
	TVWS antenna

respect to their locations, which require self-positioning function. We will also address self-positioning alternatives.

In the medium-access control (MAC) layer, we will discuss the interferences among SU networks and their coexistence based on interference power management (IPM). We will also introduce a QoS (quality of service) guaranteed dynamic spectrum assignment approach in a centralized TVWS system.

In the network layer, we will look at how to maintain the connectivity in a large TVWS network with proper routing schemes, given that the available channel is changing dynamically in the network.

Finally, in the application layer, we will discuss advanced WSDBs and novel applications based on WSDBs. Since the spectrum opportunities are time- varying, it is challenging to guarantee the QoS for users with higher demand in quality, for example, real-time requirement. We will present the WSDB with quality control (WSDB-Q) to address this problem. Besides, we will introduce various WSDBs including an area-based WSDB and an embedded broadcast WSDBs suitable for different application scenarios. A huge number of WSDs accessing WSDB also bring an opportunity for WSDB to harvest useful information such as locations and signal strengths to form the radio environment map (REM), which can be considered as a virtual sensor network without having to physically invest or maintain those "sensors." This virtual sensor network may enable many powerful services and applications that we may not even dream of previously. We will discuss it in this chapter.

Table 4.1 summarizes the technical contents addressed in this chapter in different OSI layers for ease of visualization.

In addition to the topics in Table 4.1, we will also introduce the white space devices (WSD) available in the market at the end of this chapter.

4.1 PHYSICAL LAYER

In this section, we discuss the relevant technologies in the physical layer, starting from TVWS antenna.

4.1.1 TVWS Antenna

Application layer	REM through WSD networks
	Enhanced WSDB (area-based, embedded broadcast, WSDB-Q)
Network layer	Robust TVWS networks through advanced routing
Medium access layer	Dynamic spectrum assignment
	SU networks coexistence based on interference power management (IPM)
Physical layer	Positioning
	Out-of-band leakage control
	Channel aggregation
	Spectrum identification
	TVWS antenna

TVWS spans across a very wide frequency range. If VHF band channels are included, the lowest TVWS frequency could be 54 MHz, while the highest one in UHF is 806 MHz. The signal wavelengths are from 0.37 to 5.5 m, and the relative bandwidth is larger than 174%. To design an antenna with high radiation efficiency, one arm of the radiator should be closed to a quarter wavelengths. If a dipole antenna that is independent to the ground plane is used, the size of the TVWS antenna including two arms will be around 20 cm to 2.75 m. Although a monopole antenna can rely on one radiation arm that means the radiator size could be half the size of the dipole, it needs a ground plane that could be much larger than the radiator arm is. In advanced WSD with multiple antennas such as multiple-input, multiple-output (MIMO) architecture, besides the size of signal antenna itself, enough separation among multiple antennas is also required to obtain good diversity.

To cope with large bandwidth requirement, the design ideas in the previous work [1,2] for ultra-wideband and frequency-independent antennas would be good references. A number of antenna bandwidth widening approaches can be applied directly.

Another challenge is the size of the antenna. Since the frequency of TVWS is lower than most of current wireless communication systems such as Wi-Fi, 3G, and LTE, the physical size of the antenna for this band is larger. For other bands, the consideration of deployment and the size of the hand-held device make it more desirable to have a compact antenna. To increase the spectrum efficiency, TVWS typically uses as low an emission power as possible in the transmission mode. This results in a high antenna efficiency requirement. Some concepts used in the small antenna design for broadcast receivers (e.g., sacrificing antenna efficiency to achieve small size) are not suitable here.

Here, we introduce a new method to merge the system circuit and the antenna so as to shrink the total size of the device. This can systemically reduce the size of the whole device including the antenna while keeping high transmission efficiency.

Traditionally, the antenna and the system circuit function independently. With the emergence of portable wireless devices, there is a need to arrange the whole system, including the antenna and system circuit in a compact case [3,4]. To achieve this, the antenna needs to be placed in close proximity to the system circuit board by either stacking the antenna on the circuit board or integrating the antenna and system circuit on a single printed circuit board (PCB) as shown in Figure 4.1. The designers need to handle the cross-coupling of the antenna and circuit parts with care. Note that in these cases, the antenna radiation part is usually much smaller than the system ground plane on which the system circuit board is resided [5,6].

In UHF TVWS, the antenna size could be as big as the system PCB. If we stack the antenna and the system PCB, the radiation pattern of the antenna will be significantly affected by the system circuit. If we arrange the antenna and the circuit parts in one PCB, the area of the whole PCB will be too large. Although some efforts have been made to shrink the TV band antenna size by using electric small antenna principle [7], the antenna gain is unfortunately reduced.

In this section, we describe a new design methodology called system embedded antenna (SEA), which embeds the system circuit into the antenna without sacrificing antenna performance. Through that, the total system size is significantly reduced. Meanwhile the antenna gain and radiation pattern are not affected by the surrounding circuit components. It is mostly suitable for compact or portable device.

The key idea of the method is to segment and shape the system PCB itself to the desired dimensions of the antenna in the radio frequency (RF) while maintaining the original size of the PCB and the circuit function in baseband by using RF chokes.

An ideal RF choke is one where its impedance is infinite in RF and zero in baseband. The RF choke completely blocks RF current but allows the baseband signal to pass. A practical RF choke is an inductor. The RF choke can be a standalone component or a designed inductive structure on the PCB.

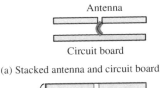

(a) Stacked antenna and circuit board

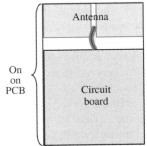

(b) Antenna and circuit in one PCB

Figure 4.1   Existing integration of antenna and system circuits.

Generally, the circuit traces (including baseband signal traces, ground (GND) plane, and DC power supply plane) can be shaped to function as the antenna radiator with suitable RF chokes. Figure 4.2 shows an example of shaping the circuit as an antenna. The whole PCB board is separated into two parts by using RF chokes, and the external DC power supply is also isolated by the RF chokes. The whole circuit board hence becomes a dipole antenna. The RF in/out of the circuit is connected to the dipole antenna feeding point. Thus, there is no need for an additional antenna.

This idea is utilized in the design of a UHF TVWS portable transceiver. Figure 4.3 shows the circuit schematic of this transceiver. The whole circuit is separated into two parts by four RF chokes: L1, L2, L3, and L4. The system is also isolated from the external power supply line in RF by L5 and L6 to avoid the effect from power line during power charged.

Figure 4.4 shows the PCB layout. The circuit board has large DC power supply and GND planes. They are A, B, C, and D. To achieve SEA, the most important components are L1 and L2. They isolate the upper halves, B and D, and the lower halves, A and C, of GND and DC power supply planes in RF, respectively. On the other hand, since the DC power supply and GND planes are arranged very close to each other (100 um, see the side view), the reactance between two planes in RF is almost zero. A and C can be considered as a single conductor in RF. Using the same principle, B and D form another single conductor in RF. Accordingly, planes A and C are the radiator part 1 (arm 1) and planes B and D are the radiator part 2 (arm 2) of a dipole antenna. This dipole antenna works in the frequency range 490–595 MHz, and all circuit parts are embedded in the antenna. The board dimensions are indicated in Figure 4.4.

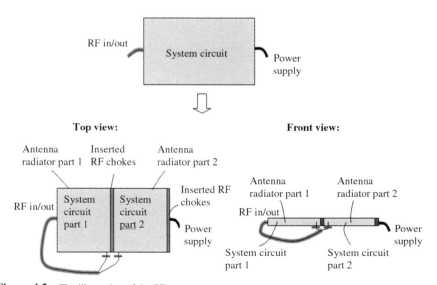

Figure 4.2   The illustration of the SEA.

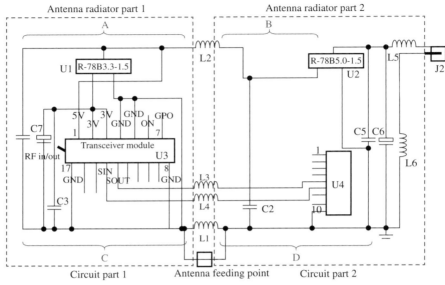

Figure 4.3 System circuit schematic of the SEA.

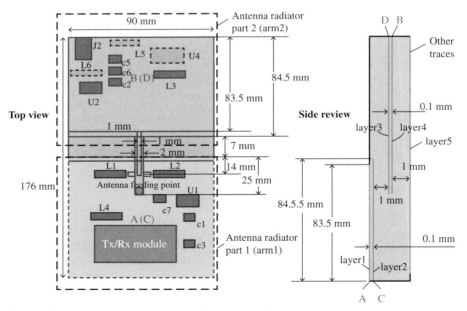

Figure 4.4 PCB layout and dimensions of the designed SEA.

L3 and L4 are used to block RF interference coupling to baseband signal traces which straddle over radiator parts 1 and 2. For other components on radiator part 1 or 2, since the huge DC power supply and GND planes are very close to each other, there is no RF voltage difference between the GND and DC power supply planes, namely each component stand on the same reference plane, RF interference will not go into these components through power supply.

Here, we select the self-resonance frequency of the RF choke as around 600 MHz. Note that the self-resonance frequency should not be much smaller than the working RF, because when the frequency is higher than the self-resonance frequency, the RF-choke's total impedance will reduce with increasing frequency due to the effect of the parasitic capacitance.

Figure 4.5 shows the RF equivalent of the PCB. It is a wideband dipole antenna with SMA connector. On the other hand, from the baseband equivalent shown in Figure 4.6, we see that the large metal planes of GND and DC power supply cover the whole board as a common PCB.

A sample SEA was fabricated and assembled for a TVWS transceiver. The assembled board is shown in Figure 4.7 and compared with the original antenna without circuits. The SEA was measured and evaluated in terms of antenna performance and transceiver performance. The antenna performance can be found in Appendix B. Here, we show the data transceiver performance with the SEA.

The TVWS communication system used for the experiment here is a wireless local area network (WLAN) with UHF converter. Its MAC-layer protocol is similar to IEEE 802.11g. However, in the PHY layer, its RF is shifted to UHF band and tunable among a few TV channels. The specific channel that the system selects depends on the availability indicated in the WSDB for the area in which the system

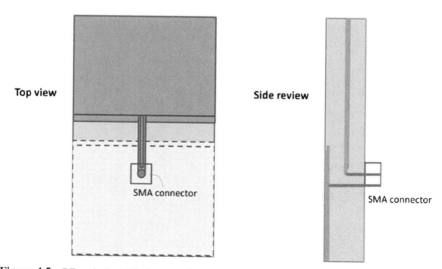

Figure 4.5 RF equivalent (dipole antenna).

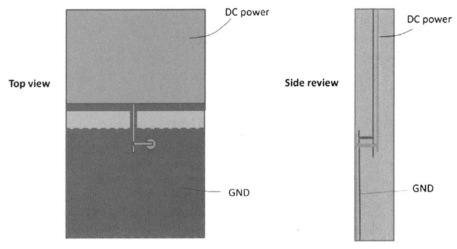

Figure 4.6 Baseband equivalent (no function change).

is working. The RF bandwidth of the TVWS system is 5 MHz, which can be included in one TV channel of 8 MHz.

The indoor performance of the TVWS SEA transceiver was compared to a transceiver having the same system circuit but using a standard 2 dBi antenna. The emission power of each transmitter was set at 20 dBm. The access point (AP) transceiver was put at a fixed location. The SEA transceiver and the common transceiver with a standard 2 dBi antenna were put at the test locations and communication links to the AP transceiver were established one after another. The communication performance was compared in Table 4.2 in terms of the link quality, signal strength,

 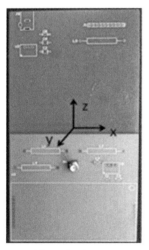

Figure 4.7 Fabricated TVWS SEA and its original antenna without circuit.

Table 4.2   Measurement Result of the Transceiver Performance of the SEA Compared with that Using 2 dBi Standard Antenna

Distance (m)	Link quality (out of 70)		Receiving signal strength (dBm)		Packet loss rate (%)	
	SEA	Standard	SEA	Standard	SEA	Standard
6	49	53	−43	−45	0	0
11	42	48	−53	−50	0	0
25	25	28	−69	−71	0	0

and packet loss rate. We can see that in the test distances, the compared parameters are quite similar. This verifies the efficacy of the SEA.

4.1.2   Spectrum Identification

Application layer	REM through WSD networks
	Enhanced WSDB (area-based, embedded broadcast, WSDB-Q)
Network layer	Robust TVWS networks through advanced routing
Medium access layer	dynamic spectrum assignment
	SU networks coexistence based on IPM
Physical layer	Positioning
	Out-of-band leakage control
	Channel aggregation
	Spectrum identification
	TVWS antenna

Spectrum identification is the core feature that differentiates CR and TVWS with other conventional communication systems. In this section, we introduce typical spectrum identification approaches for CR and TVWS, which include spectrum sensing, beaconing, geolocation database, and hybrid solution.

4.1.2.1 Spectrum Sensing

There are two major tasks within the spectrum sensing approach [8]. One is for a SU to detect PU frequencies and transmissions so as to find a spectrum hole for its own transmission without interfering the PUs. This can be viewed as finding spectrum opportunities. The other is to detect and avoid interference from other SUs so as for the SU to improve its own performance.

Transmitter Detection   Spectrum sensing consists of individual sensing and cooperative sensing.

a. *Individual Sensing.* The receiver carries out sensing independently.

The individual sensing approaches can be classified into energy detection, cyclostationarity-based sensing, radioidentification-based sensing, waveform-based sensing, and matched filtering [9].

The energy detector has a simple structure. It senses a signal by comparing the output of the energy detector with a predetermined noise threshold. It is easy to be implemented as it does not need to know the characteristics of the signal to be detected. However, the probability of error in detection is high due to noise uncertainty.

Cyclostationary features are wide sense stationary signals with no correlation. They are used in the cyclostationary feature sensing method to distinguish a signal from noise. The cyclostationary feature detector has better performance in terms of robustness to noise uncertainty and detection sensitivity.

If the waveform of the desired signal is known, the detector can carry out the autocorrelation between the received signal and the stored template. This is a waveform-based sensing. This kind of sensing outperforms energy detection in terms of reliability, and its performance continuously improves along with increasing the length of known signal template.

There is another feature-based sensing technique, which extracts selected features of the received signal such as channel bandwidth and central frequency for sensing decision through various classifications. This is called radioidentification-based sensing.

Matched filtering requires full knowledge of the detected signal, including bandwidth, operating frequency, modulation, waveform, and even frame format to demodulate the signal for sensing decision making [9]. It is clear that the matched filter approach is more reliable but complicated.

Figure 4.8 indicates the performance versus complexity of individual spectrum sensing approaches.

There are some technical challenges using individual spectrum sensing approaches. One of the main challenges is channel uncertainty. A practical wireless channel consists of multiple paths. The multipath is usually time-varying due to the movements of surrounding objects or the transceiver. This causes fading and blocking that results in drastic changes in the received signal strength. The receiver could incorrectly detect the PU

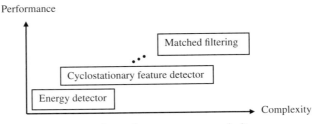

Figure 4.8   Individual spectrum sensing performance versus complexity.

transmissions due to fading and blocking. A typical problem is the "hidden node" problem, which means there is a blockage between the SU receiver carrying out sensing and the PU transmitter, but there is no blockage between the PU transmitter and PU receivers [10]. In this case, spectrum sensing cannot effectively achieve the purpose of detecting PU transmission. The second challenge is the uncertainty due to aggregated interference. When a number of SUs work in the same area in an uncoordinated manner, the mutual interference of SUs may affect their sensing performance. In other words, if every SU system operates independently, all other SU systems' signals could degrade one SU's spectrum sensing accuracy, and this effect is time-varying. This could lead to missed detection of PU signals. The third challenge is the noise uncertainty. For the best detection performance, especially for weak PU transmission detection, the noise power spectrum density (PSF) should be known. However, in a practical system, the noise floor is actually estimated by a SU receiver. This estimation accuracy is affected by the device's time-varying thermal noise due to temperature drift.

Another challenge with the individual sensing is that a WSD may need to sense extremely weak TV signals to overcome the "hidden node" problem (false negatives), but at the same time this leads to overdetect the TV signals very far from the SU perhaps hundreds of kilometres away (false positives). This may remove most portions of usable white spaces. Very recent studies in the United States shows that a threshold of −114 dBm (the conservative level given by the regulator to protect PUs) reduces the recoverable white spaces by a factor of three. In the United Kingdom, initial modeling studies performed at BT show that a cognitive device with this sensing sensitivity level will consider all DTT channels as occupied, and hence have no available frequency for TVWS operation if it relies on native sensing only [10].

b. *Cooperative Sensing*. To increase reliability, detection of protected transmitters can be carried out through information exchange among a number of SUs. This can significantly decrease the probability of missed detection caused by fading and blocking. The effect of noise uncertainty can also be mitigated to some extent. This is called cooperative sensing. Cooperative sensing needs sensing information exchanged among many SUs. The information exchange in cooperative detection can be done through either centralized networks or distributed networks.

Cooperative sensing requires additional communications links among multiple SU nodes to be established before detecting the available frequency. This may not be always feasible. The performance of the cooperative sensing also largely depends on the number of cooperating nodes and their spatial arrangement. This results in difficulty to test and certify the detection quality [10]. Another purpose of spectrum sensing is to find cleaner channels in available TVWS channels to guarantee the communication quality of the CR. In this case, the detection usually focuses on measuring the noise levels in the given channels. It does not require the same accuracy level as that of PU detection.

4.1.2.2 Beaconing

Beaconing is a technique in which a spectrum user intentionally announces its frequency usage information publicly on a common channel or on its own frequency using common protocol and format. Through this approach, the other radios can easily detect this spectrum user and avoid interfering with it. Beaconing has been widely used to coordinate multiple access points (APs), and also for clients to associate to an AP in wireless local area networks WLAN.

In CR networks, beaconing can be used in a few ways [11].

Single-Transmitter Beaconing. A PU can broadcast its frequency information using an easily detected signal in a standardized format so that SUs can reliably detect the PUs. Since the beacon signal does not use complex and high-order modulation, the required demodulation signal-to-noise ratio (SNR) is low. Thus, the beacon information can be easily detected and demodulated. This increases spectrum sensing robustness. However, it occupies addition radio resource, and may require modifications to the PUs that are already in operation that is practically difficult if not impossible.

Multiple-Transmitter Beaconing: Transmitter beaconing is not limited to each transmitter broadcasting only its own information. A transmitter can also broadcast beacon signals for a set of frequencies and locations that are occupied by PUs. We will describe this scheme with more details later.

Receiver Beaconing. The reason why the transmission of the PU has to be protected is to prevent the PU receiver from being polluted by the stronger signals of other SUs. If receiver's location and its working frequency are known, interference can be effectively avoided. Thus, the beacon signal can contain the information of receiver's location and frequency. Of course, if the PU transmitter knows its receivers' locations, the easiest way would be for the PU transmitter to broadcast its receivers' locations and receiving frequency. It is also possible for the receiver to intentionally emit the service frequency on another channel.

4.1.2.3 White Space Database

A WSDB is a central server that records the information of all PUs' locations and protection contours so as to provide frequency availability (white space) at different locations. The WSDB is periodically updated. The WSDs have to look up the database through a standard interface to find out the available frequencies at its own position prior to its transmission. In a basic WSDB, the frequency availability is calculated only based on PUs' protection contours. The basic function is illustrated in Figure 4.9.

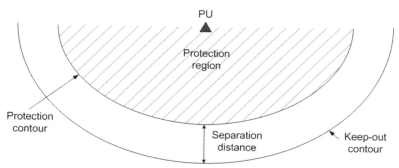

Figure 4.9 Protection region, protection contour, seperation distance, and keep-out contour of a primary user.

The WSDB will create a protection region around a PU based on the PU's operating information. The protection region is basically the coverage of the PU's transmitter where its transmit signal power is above a predetermined threshold. The transmit signal power level at different locations is computed using propagation models that are standardized across all WSDB operators [12]. To protect the PU's receivers inside the protection region, a separation distance from the boundary of the protection region is used to form a keep-out contour as shown in Figure 4.9. A channel is available to a SU when it is located outside of the keep-out contour. The WSDB will determine whether the SU is within or outside of some or all the PUs' keep-out contours and then informs the SU of its available channels.

Advanced WSDBs could provide additional information on the white space quality according to the number of active WSDs and their powers in the related area. To achieve that, the WSDB requires the WSDs to feedback their radio information, including location, transmission frequency and power, RF bandwidth, and so on.

The WSDB approach completely eliminates the various uncertainty issues in estimating PU's transmission. Thus, it is the safest approach to protect PUs. However, it also has a few disadvantages. First, an additional communication channel is required between the WSD and the WSDB. The additional channel is usually through wired internet. This raises a question: why would we need the WSD communication in white space if we already have an internet connection? Second, since the WSDB's updating period is relative long, the change of PUs' activities may not be reflected in the WSDB instantly. This results in precious spectrum resources being left idle and thus the spectrum utilization rate achieved is not as high as that based on other real-time approaches.

As mentioned, TVWS are the first frequency bands opened up for practical CR application. To minimize the risk of interfering with the PUs, regulators have decided to use WSDBs to coordinate the PUs and the WSDs. This decision may be conservative but it gives certainty for broadcaster, wireless microphone, and other PUs in the TV bands.

4.1.2.4 *Hybrid Solution*

Spectrum identification can also be a hybrid solution. To ensure PUs' quality in a reliable manner, it is necessary to use a WSDB. This is the first level protection, which only coordinates the PUs and the SUs. Among SUs, spectrum sensing technology can be used to avoid mutual interference so as to enhance the coexistence efficiency in a shared white space.

With the aid of spectrum sensing, the requirements on WSDB, for example, the protection contours, could also be potentially relaxed. This will greatly enhance the spectrum utilization. Spectrum sensing may also help to adjust the various parameters used in a typical WSDB. On the other hand, with WSDB, the requirements on spectrum sensing of PUs can be lowered as well as to make spectrum sensing implementable in a cost-effective manner.

The result of spectrum identification may support other functions in the different layers of the cognitive radio networks (CRN). For example, with WSDB, there are more enhanced functions such as area-based protection, embedded broadcast WSDB, and revenue optimization with HPC. We will discuss these enhanced features in the application layer in Section 4.4.

4.1.3 Channel Aggregation

Application layer	REM through WSD networks
	Enhanced WSDB (Area-based, embedded broadcast, WSDB-Q)
Network layer	Robust TVWS networks through advanced routing
Medium access layer	Dynamic spectrum assignment
	SU networks coexistence based on IPM
Physical layer	Positioning
	Out-of-band leakage control
	Channel aggregation
	Spectrum identification
	TVWS antenna

TVWS includes a plentiful of spectrum opportunities that could support high-throughput wireless transmissions. However, these available frequencies are usually noncontiguous in spectrum and time-varying, which implies that a WSD has to quickly reconfigure itself to aggregate multiple spectrum fragments. Simultaneously, using multiple noncontiguous channels for communication is termed as channel aggregation.

One technique suitable for this requirement is a variant of OFDM called non-contiguous OFDM (NC-OFDM) [13]. NC-OFDM is capable of deactivating some of its subcarriers across its transmission bandwidth where the PUs are occupying or when interference is strong. NC-OFDM is illustrated in Figure 4.10. Figure 4.10a shows that NC-OFDM has huge working frequency range. To avoid interfering with the PU, it does not tune the carrier in RF. Instead, it just switches off a few unwanted subcarriers. It can also avoid heavy interference in the same way. To do

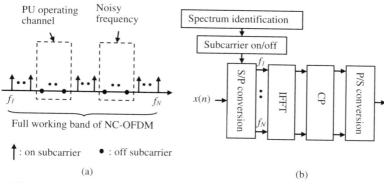

Figure 4.10 Illustration of NC-OFDM.

so, the transmitter needs to acquire the spectrum information from the spectrum identification introduced earlier and disable and enable some branches of the serial to parallel (S/P) converter correspondingly through the subcarrier on/off controller, as shown in Figure 4.10b. The receiver performs the reverse operation of the transmitter after knowing the subcarrier on/off information.

Another possible technology for channel aggregation is frequency band multicarrier (FBMC). It can also easily control subcarrier switching on and off to avoid the PUs and interference. FBMC has another advantage of good OOB leakage control. It will be introduced in the next subsection.

4.1.4 Out-Of-Band Leakage Control

Application layer	REM through WSD networks
	Enhanced WSDB (area-based, embedded broadcast, WSDB-Q)
Network layer	Robust TVWS networks through advanced routing
Medium access layer	Dynamic spectrum assignment
	SU networks coexistence based on IPM
Physical Layer	Positioning
	Out-of-band leakage control
	Channel aggregation
	Spectrum identification
	TVWS antenna

To protect PUs, regulators defined a stringent OOB leakage limit for WSDs working in TVWS bands. For example, in FCC's TVWS regulation, a portable WSD has 16 dBm nominal emission power if it works in adjacent channels of a PU, its OOB leakage should be smaller or equal to −56.8 dBm per 100 KHz, which means that the in-band to OOB PSD ratio is around 55.4 dB. This requirement is much higher than most of other existing wireless systems such as Wi-Fi and cellular system. How to design TVWS in a cost-effective manner is a challenge.

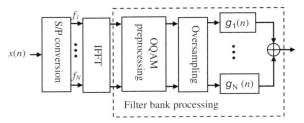

Figure 4.11 System block diagram of FBMC modulator.

Traditionally, filters are used to filter off unwanted signals such as OOB. It is necessary to use additional filtering if a conventional modulation such as OFDM or direct spreading spectrum is adopted for WSDs. RF OOB filtering at a target frequency is not feasible, because a WSD needs to frequently tune its working frequency to avoid interfering with the PUs, and hence its RF frequency is not fixed. A tunable RF filter with sharp roll-off is too expensive. Its performance is also unstable along with temperature and frequency changes.

In contrast, coping with OOB problem in baseband is more practical and robust. A straightforward approach is to carry out digital OOB filtering after conventional modulation, for example, filtered OFDM [14]. However, this results in considerable additional computation to meet the requirement.

Alternatively, we can use new modulations with better adjacent channel leakage ratio (ACLR) performance. A promising modulation is FBMC [15]. The block diagram of an FBMC modulator is shown in Figure 4.11. Both FBMC and OFDM are based on multi-carrier modulation. The main difference between FBMC and OFDM is that FBMC replaces cyclic prefix (CP) of OFDM by a filter bank processing, where $g_k(n)$ denotes the kth filter. By eliminating CP, FBMC achieves higher spectrum efficiency. Moreover, the filter bank delivers much better OOB performance than OFDM.

In [15], OOB performance of FBMC, filtered OFDM, and conventional OFDM is compared. As shown in Figure 4.12, FBMC has the best OOB performance that meets 55.4 dB in-band to OOB PSD ratio required by the regulator.

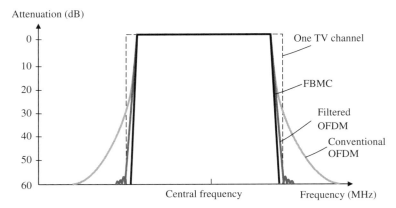

Figure 4.12 Comparison of OOB performance.

Besides, there are other approaches such as generalized frequency division multiplexing (GFDM) [16] for OOB leakage reduction. GFDM uses flexible pulse shaping to subcarrier waveforms in the modulator to reduce the OOB radiation to the adjacent channels. However, this also introduces intersubcarrier interferences and degrades the performance of GFDM reception. Thus, in a GFDM receiver, a successive interference canceller is needed to cancel the self-inter-carrier-interference.

4.1.5 Positioning

Application layer	REM through WSD networks
	Enhanced WSDB (area-based, embedded broadcast, WSDB-Q)
Network layer	Robust TVWS networks through advanced routing
Medium access layer	Dynamic spectrum assignment
	SU networks coexistence based on IPM
Physical layer	*Positioning*
	Out-of-band leakage control
	Channel aggregation
	Spectrum identification
	TVWS antenna

To avoid interfering PUs, current TVWS regulations require the WSDs with fixed locations or mode II portable devices to periodically access the WSDB for spectrum opportunities in terms of their locations. This means the WSDs except mode I devices need to be aware of and provide their own locations when querying the WSDB.

In an outdoor environment, GPS works most of the time except for some highly urbanized areas. A WSD adopting GPS for positioning is currently assumed to be the norm. To solve the issues of poor positioning accuracy in urban jungles and perhaps even in indoor environments, alternative positioning methods are sought for.

Cellular wireless or Wi-Fi could provide self-positioning to some levels of accuracies. This requires that the WSD equips with cellular or Wi-Fi transceiver. This option is cost-effective only if the WSD equipment comes with built-in cellular or Wi-Fi for other purposes.

On the other hand, WSD operates in the TV band using the same frequency bands as digital TV (DTV). DTV signals could be exploited for WSD self-positioning without much added cost. Since TV band signals have lower frequencies than GPS, its propagation loss is lesser, which can result in a power advantage of more than 40 dB over GPS signals.[1,2] TV band signals also have better penetration

---

[1] Typical GPS satellite operates at 1575 MHz L1 frequency. The distance from satellites to earth is about 21 000 km.

[2] http://gpsinformation.net/main/gpspower.htm.

through buildings and walls. Thus, TV signals are more suitable for indoor positioning as well as outdoor.

Using DTV signals for positioning has been reported in a number of existing works. The basic idea for this kind of self-positioning is to use the time differences of arrivals of the signals from multiple synchronized TV towers to calculate the WSD's own location as illustrated in Figure 4.13. According to the principle of time-of-difference-of-arrival (TDOA) 2D trilateration positioning [17], the coordinates (x, y) of the WSD is the solution to the Equation Group 4.1.

$$
\begin{cases}
\Delta_1 = t_1 - t_2 = \dfrac{\sqrt{(x_1 - x)^2 + (y_1 - y)^2}}{c} - \dfrac{\sqrt{(x_2 - x)^2 + (y_2 - y)^2}}{c} \\[2ex]
\Delta_2 = t_2 - t_3 = \dfrac{\sqrt{(x_2 - x)^2 + (y_2 - y)^2}}{c} - \dfrac{\sqrt{(x_3 - x)^2 + (y_3 - y)^2}}{c} \\[2ex]
\Delta_3 = t_3 - t_1 = \dfrac{\sqrt{(x_3 - x)^2 + (y_3 - y)^2}}{c} - \dfrac{\sqrt{(x_1 - x)^2 + (y_1 - y)^2}}{c}
\end{cases} \tag{4.1}
$$

where Δ_1, Δ_2, and Δ_3 are the time differences of arrivals measured in the WSD between TV towers 1 and 2, 2 and 3, and 3 and 1, respectively; and c is the speed of light. The positioning accuracy can be improved by increasing the number of the TV towers. This is called multilateration positioning.

In [18], a positioning system that makes use of the ATSC DTV synchronization signals is proposed. In [19], the authors propose to provide positioning using the transmitter identification watermark signals, which are pseudorandom sequences embedded into normal ATSC DTV signals. ATSC DTV systems are mainly deployed in North America, such as the United States and Canada. In [20], a positioning system that uses the time-domain synchronous OFDM signals as specified in the Chinese DTV standard is proposed. This standard is mainly deployed in China and its territories. In [21], the authors use the known pilot signals in Digital Video Broadcasting—Terrestrial (DVB-T) system for channel estimation. However, in single-frequency DVB-T systems, the estimated channel is a composite channel from multiple transmitters. In [22], a positioning system is proposed by exploring the DVB-T2 transmitter signature signals. With transmitter-dependent signature

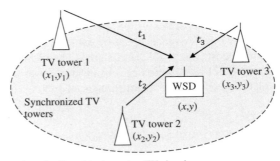

Figure 4.13   Illustration of self-positioning using TV signal.

waveforms, signals transmitted from the different transmitters can be easily identified and the corresponding channel response can be measured individually. In addition, the length-2^{16} signature waveforms contribute a SNR gain of around 48.2 dB after correlation. Hence, the channel response and timing estimation will be more reliable than that obtained based on DVB-T pilot signals. The positioning system in [22] uses a threshold-based timing estimation approach in which the threshold is computed to maximize the probability of correctly detecting the timing of each transmitter. The computations of thresholds require the knowledge of the channel propagation loss from each transmitter, which may not be obtained easily in practical applications. In the absence of such knowledge, the performance of the proposed system degrades significantly.

To overcome the drawbacks of the existing approaches, two methods based on DVB-T2 signals are presented in this section. One is threshold-based approach, which is applicable to small-size SFNs. This threshold-based approach differs from the existing methods in the sense that the threshold is not computed using the knowledge of the channel propagation loss that cannot be obtained easily. Instead, the threshold is calculated based on the noise power. Thus, it is more robust to different signal-to-noise ratios. The other approach is based on iterative timing estimation that can be applied to both small-size and large-size SFNs. It iteratively extracts the multipath elements one by one, and hence is more suitable for the environment with heavy reflections. It is also suitable for large SFNs, where multiple tower signatures, same as the local ones from other zones, may be received. It is shown that the positioning accuracy can meet ± 20 m with a probability of 95% if the signal-to-noise ratio is 20 dB. More details can be found in Appendix C.

4.2 MEDIUM ACCESS CONTROL LAYER

The basic methods of sharing medium along multiple wireless users are time division multiple access (TDMA), frequency-division multiple access (FDMA), space-division multiple access (SDMA), and CDMA. The main issue that needs to be considered specifically for TVWS communications is that a large number of various SU wireless systems using different bandwidths, modulations, transmission powers, and signaling formats may occupy the same band in an opportunistic manner. According to the regulations, all these SU systems have the option to access any frequency from the list of available frequencies for their TVWS operation. Since there is likelihood of overlapping with each other in terms of frequency, each of SU system should cognize the operations of the others and reconfigures itself for its own operation with minimal interference to others or being interfered by other systems. For instance, it will be important for IEEE 802.11af networks to detect the presence of nearby IEEE 802.22 networks, which may impose serious interference. The IEEE 802.11af network will then have to set up its own parameters properly to avoid network capacity drop. The same applies to the IEEE 802.22 systems. In other words, various SU networks are CRNs coexisting in the same bands.

Before we discuss the coexistence of SU networks, we summarize the major differences between the systems competing for the same bands:

1. *Bandwidth.* Operating channel bandwidths may not be identical for different networks. For example, IEEE 802.22 uses one TV channel (6, 7, or 8 MHz) as the operating bandwidth while IEEE 802.15.4m may use a small fraction of a TV channel.

2. *Transmission Power.* We know that in regulation viewpoint, a fixed WSD can transmit up to 36 dBm EIRP, while a portable WSD can transmit only up to 20 dBm EIRP. Transmission power also varies between different standards. In IEEE 802.22, the station is allowed to transmit up to 36 dBm EIRP, while in IEEE 801.11af, the transmit power can only be up to 20 dBm.

3. *Signaling.* Each system uses its own protocol to coordinate transmitters and receivers. The signaling signals are usually different between different systems.

4. *Other Signal Characteristics.* In addition to bandwidth and transmission power, the other signal characteristics such as modulation, the number of subcarriers, and the guard interval are typically different as well.

There are basically two ways to coordinate various SU systems, that is, distributed or centralized such as IEEE802.19.1.

In distributed CRNs, the various systems rely on individual detections to obtain suitable channels for transmission. There is no coordination between these systems. This may work when the total number of white space users is small. However, when the TVWS grows up and the quantity of SU networks increases, purely distributed CRNs may face low efficiency problem due to mutual interference between SU networks. The problem is especially severe for TVWS system due to the presence of high-power SU devices and the strong propagation nature of TV bands. To fully cognize the other systems with different parameters introduced above, a simple spectrum sensing technology may not be sufficient. We need a comprehensive way to know the parameters of the other systems, for example, through a common channel and common format to share system parameters information. Moreover, individual adaptations would not achieve the global optimization of the whole TVWS spectrum.

The other way is to coordinate SU networks through a centralized server. This can be understood as an extension of the WSDB. The coordination among SU networks is not as strict as the protection for PUs. Instead, the extended WSDB will guide SU networks with information of other systems, interference level, and so on. For SU networks with difference technologies, the centralized server will admit them in terms of IPM [23]. With the coordination by the centralized server, the total utilization of TVWS spectrum could be improved.

This is similar to the traffic of vehicles. When cars were just invented, there were few vehicles on the road. It would be good enough to rely on the drivers themselves to detect the other vehicles and then change lanes or cross junctions accordingly. After the quantity of vehicles increases, we then need road signs and traffic lights to help drivers to use the road better.

In this section, we introduce two approaches for SU network coexistence with different protocols. One is based on the interference power management and the other is based on the dynamic spectrum assignment under optimization theory. Both are based on centralized approach.

4.2.1 Secondary User Networks Coexistence Based on IPM

Application layer	REM through WSD networks
	Enhanced WSDB (area-based, embedded broadcast, WSDB-Q)
Network layer	Robust TVWS networks through advanced routing
Medium access layer	Dynamic spectrum assignment
	SU networks coexistence based on IPM
Physical layer	Positioning
	Out-of-band leakage control
	Channel aggregation
	Spectrum identification
	TVWS antenna

In this subsection, the secondary user coexistence based on the IPM [23] is discussed. We will also describe channel ordering based on interference power and propagation models, respectively, followed by numerical results.

4.2.1.1 IPM Through WSDB

The WSDB uses IPM by estimating the interference powers among WSDs and allocates channels to WSDs such that they do not cause excessive interference to each other. We provide the computations of the interference powers among the WSDs and the interference power constraints that the WSDB will use to determine whether a channel is available for a WSD.

We denote every WSD with an index number j. J_k means the set of current operating WSDs in channel k. The index number 0 means a new requesting WSD. It is supposed that these WSDs are based on difference protocol and cannot self-coexistence. If the WSDB permits WSD 0 to operate in channel k, the interference plus noise power experienced by a WSD j that has already been operating in channel k is given as

$$I_j^k = \sum_{\substack{i \in J_k \\ i \neq j}} g_{ij}^k P_i^k + \sum_{\substack{\varepsilon = -L \\ \varepsilon \neq 0}}^{L} \sum_{i \in J_{k+\varepsilon}} g_{ij}^k P_{i,OOB}^{\varepsilon,k} + g_0^k P_0^k + T_j^k + N_j, \forall_j \in J_k,$$

(4.2)

where g_{ij}^k is the channel gain from WSD i to WSD j in channel k; P_i^k is the transmit power of WSD i in channel k; $P_{i,OOB}^{\varepsilon,k}$ is the out-of-band (OOB) power of WSD i in channel k, which is operating at ε channels away from channel k; T_j^k is the total interference power from the PUs to WSD j in channel k; and N_j is the noise power at WSD j. In Equation 4.2, the interferences from WSDs in adjacent L channels are included based on their OOB power emission.

Similarly, if the WSDB permits WSD 0 to operate in channel k, the interference-plus-noise power experienced by a WSD j that has been operating at l channel away from channel k is given as

$$
\begin{aligned}
I_j^{k+l} = \sum_{\substack{i \in J_{k+1} \\ i \neq j}} g_{ij}^{k+l} P_i^{k+l} + \sum_{\substack{\varepsilon = -L+l \\ \varepsilon \neq 0}}^{L+1} \sum_{i \in J_{k+\varepsilon}} g_{ij}^{k+l} P_{i,OOB}^{\varepsilon,k+l} + g_0^{k+l} P_{0,OOB}^{l,k+l} \\
+ T_j^{k+l} + N_j, \forall_j \in J_{k+l}.
\end{aligned}
$$

(4.3)

The current operating WSDs in both co-channel and adjacent channels must be able to withstand the additional interference generated from the requesting WSD if it is to operate in the channel. Hence, the constraints to determine whether a channel k is available to the requesting WSD are given as

$$
I_j^{k+l} \leq \bar{I}_j, \forall_j \in J_{k+l}, \; l = -L, \ldots, 0, \ldots, L,
$$

(4.4)

where $\bar{I}_j$ is the interference-plus-noise power threshold that WSD j can tolerate.

When a requesting WSD queries the WSDB, the WSDB needs to determine

$$
g_{0j}^k P_0^k \leq \bar{\gamma}_j, \; \forall_j \in J_k,
$$

(4.5)

and

$$
g_0^{k+l} P_{0,OOB}^{l,k+l} \leq \bar{\gamma}_j, \forall_j \in J_{k+l}, \; l = -L, \ldots, 0, \ldots, L,
$$

(4.6)

where

$$
\begin{aligned}
\bar{\gamma}_j = \bar{I}_j - \sum_{\substack{i \in J_{k+1} \\ i \neq j}} g_{ij}^{k+l} P_i^{k+l} - \sum_{\substack{\varepsilon = -L+l \\ \varepsilon \neq 0}}^{L+1} \sum_{i \in J_{k+\varepsilon}} g_{ij}^{k+l} P_{i,OOB}^{\varepsilon,k+l} - T_j^{k+l} - N_j, \forall_j \in J_{k+l}, \\
l = -L, \ldots, 0, \ldots, L
\end{aligned}
$$

(4.7)

is the interference margin for WSD j which can be precomputed. If a channel is able to satisfy all the constraints in Equations 4.5 and 4.6, the WSDB will approve the channel for the requesting WSD to use. When the requesting WSD decides to operate in a channel and informs the WSDB, the WSDB can compute the new interference margin for all the affected WSDs by simply subtracting the left terms in Equations 4.5 and 4.6 from their current respective interference margin. Therefore, there is no need to recompute the interference margins from Equation 4.7 whenever a requesting WSD joins the unlicensed spectrum.

4.2.1.2 *Channel Ordering Based On Interference Power*

When applying the IPM scheme, the information used to compute Equations 4.5–4.7 to determine channel availability can also be used to estimate the total interference power that the existing WSDs generate to the requesting WSD. The estimated total interference power generated by the existing WSDs to the requesting WSD in each channel can be computed by the WSDB using the equation given by

$$I_0^k = \sum_{i \in J_k} g_{i0}^k P_i^k + \sum_{\substack{\varepsilon = -L \\ \varepsilon \neq 0}}^{L} \sum_{i \in J_{k+\varepsilon}} g_{i0}^k P_{i,OOB}^{\varepsilon,k} + T_0^k + N_0. \qquad (4.8)$$

The WSDB can order the channel quality for the requesting WSD from the channel with the smallest I_0^k to the largest I_0^k as an addition service to the requesting WSDs.

Channel gain information between WSDs is required for the WSDB to estimate the interference among WSDs and determine the channels available to WSDs under the IPM scheme. However, it is hard for the WSDB to know the exact channel gains of WSDs. This is similar to the case where it is hard for the WSDB to know the channel gains from TV transmit towers to TV receivers that are needed to form the protection contours of the PUs. Hence, WSDB uses propagation models for channel gains to determine the protection contours. In this IPM scheme, propagation models are used for the channel gains of WSDs as well.

A common propagation model used for WSDs is the Hata-Okumura model. The path loss of Hata-Okumura model is given as

$$\begin{aligned} PL_H = &\, 26.16 \log f_c - 13.82 \log h_t - 3.2(\log(11.75 h_r))^2 \\ &+ (44.9 - 6.55 \log(h_t)) \log d + 74.52, \end{aligned} \qquad (4.9)$$

where fc is the frequency in MHz, d is the distance in km, and ht and hr are the antenna heights of the transmitter and receiver in meter, respectively.

The FCC uses it to determine the minimum separation distances required for WSDs to be away from protection contours. However, the Hata-Okumura model used for predicting field strength levels from transmitter that has a height of 30–200 m and may not be a good propagation model for WSDs with low antenna height. Therefore, the FCC has suggested to use OET TM 91-1 model for WSDs with antenna height that is less than 30 m [24]. The OET TM 91-1 model is originally developed for predicting field strength levels within 10 miles of a transmitter with a height of less than 300 ft and operate in frequencies between 40 and 1000 MHz [25]. Therefore, we use the OET TM 91-1 model for WSDs that have low antenna height. Assuming that the antenna gain is 0 dBi, the pathloss of OET TM 91-1 model is given as

$$PL_{91-1} = 20 \log f_c - 20 \log(h_t h_r) + 40 \log d + 87.97. \qquad (4.10)$$

4.2.1.3 *Simulations*

In the simulations, we assume that a WSD will appear randomly distributed over a $10 \times 10 \, \text{km}^2$ area and queries the WSDB for a channel to operate. There are five

TVWS channels: 634, 642, 650, 658, and 666 MHz with each channel having a bandwidth of 8 MHz. The WSDB will determine if the five channels are available to the requesting WSDs based on the IPM scheme and also perform channel ordering of the available channels. The ratios of OOB-to-inband power spectral density (PSD) of a WSD in all simulations are 55.4 dB. When a requesting WSD receives the available channels, it will choose the channel that has the lowest I_0^k and that it is below a threshold. If two or more channels have the same lowest I_0^k, then the WSD will always choose the lowest frequency in our simulations. We assume the WSD has the noise floor of −96 dBm and its sensitivity is −80 dBm. After the requesting WSD selects a channel to operate on, it will report back to the WSDB on the channel it operates and the interference margin it can tolerate. These will enable the WSDB to update its information in the IPM system and also includes the requesting WSD into the operating WSDs list in its database.

The simulation results are shown in Figure 4.14. The performances with IPM and channel ordering are compared to that without IPM and channel ordering (non-coexistence), where as long as the WSDs do not interfere with the PUs, the WSDB will approve the WSDs to use the channel. "NC" represents noncoexistence. The results are averaged over 100 Monte Carlo simulations.

From the results, it can be seen that when the index of the requesting WSDs is small, most of the requesting WSDs are able to obtain an available channel under both IPM and noncoexistence schemes. However, when the index of requesting

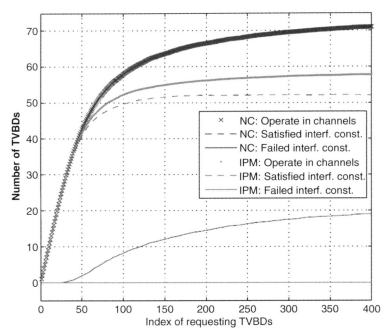

Figure 4.14   Performances of interference power management coexistence scheme and noncoexistence scheme when all the WSDs' locations are known.

WSDs gets higher, the increase in the number of WSDs that are able to obtain an available channel slowed down. This is because the interferences among the WSDs and to PU increased. The new requesting WSDs could hardly find any channel that has the interference plus noise power that is lower than its threshold to operate in.

Although under the noncoexistence scheme, more WSDs are able to find a channel to operate, there is a chance that a new WSD causes other operating WSDs to fail. Hence, the overall throughput of the whole system does not increase. In fact, if one new WSD causes more than one operating WSDs to fail, it actually decrease the overall throughput of the system. The term satisfied interference constant ("satisfied interf. const." in Figure 4.14) means the overall throughput. For IPM scheme, the "operate in channels" is equal to the overall throughout, because there is no operating WSD failing due to another new WSD's admission. In contrast, the overall throughput of noncoexistence scheme is smaller than its "operate in channels."

We can see that the overall throughput under IPM is larger than that without IPM when the number of WSDs trying to use TVWS is large. The larger the number of WSDs, the better the IPM scheme compared to noncoexistence scheme.

Note that we have simulated the IPM performance in the case that WSDs have equal bandwidth. More advanced IPM should be able to deal with variable bandwidth WSDs. This can be extended based on the equations introduced in this subsection. Another centralized TVWS management approach capable of handling WSDs with variable bandwidth is called dynamic spectrum assignment and will be presented in the next subsection.

4.2.2 Dynamic Spectrum Assignment

Application layer	REM through WSD networks
	Enhanced WSDB (area-based, embedded broadcast, WSDB-Q)
Network layer	Robust TVWS networks through advanced routing
Medium access layer	*Dynamic spectrum assignment*
	SU networks coexistence based on IPM
Physical layer	Positioning
	Out-of-band leakage control
	Channel aggregation
	Spectrum identification
	TVWS antenna

In the dynamic spectrum assignment scheme, the available TVWS and requests for spectrum from WSDs are denoted as spectrum fragments. An illustration of such spectrum fragments is shown in Figure 4.15. Due to the existences of wireless microphone users and other non-TV users, the original available TVWS spectrum fragment may not be always an integer number of TV channels. In addition, WSDs that want to access those available spectrum fragments can also have heterogeneous bandwidth requirements due to different applications they are carrying or different access technologies they are using.

Figure 4.15 An illustration of the available spectrum fragments.

In [26], the above problem is modeled to maximize the overall spectrum utilization as a multiple knapsack problem [27]. An exact solution and a heuristic solution are provided. In [28–30], a few different performance metrics are considered such as to minimize spectrum fragmentation and maximize profit. A backtracking algorithm is proposed to search for the best allocation. In [31] and [32], the neighboring spectrum fragments can be aggregated if they are within the spectrum aggregation range imposed by hardware. The associate spectrum assignment problems are investigated. These works assume that the spectrum requirements from the WSDs are revealed at the same time. This corresponds to dedicating a time period to collect the requests from WSDs and optimizing the resource with complete knowledge of requests.

Let us consider a TVWS network in which there is a centralized spectrum manager (SM) to control the access of the WSDs within its coverage region as shown in Figure 4.16. The SM is connected to the WSDB to periodically obtain the available spectrum fragments, which will then be assigned to WSDs requesting spectrum access. The available spectrum fragments as well as the requested bandwidth from WSDs can be of different bandwidth. It is assumed that WSDs request bandwidth in a sequential manner and the SM is able to know the distribution of bandwidth requests through past observations. Upon the arrival of a bandwidth request, the SM has to determine which fragment the request should be assigned to in terms of maximizing the spectrum utilization rate.

The above problem has been formulated as a stochastic sequential decision-making problem. The optimal spectrum assignment policy to maximize the overall spectrum utilization of the TVWS is computed through a value iteration method as shown in Algorithm 4.1.

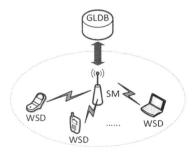

Figure 4.16 System model of spectrum manager.

Algorithm 4.1

Algorithm for computing $J^*(s), s \in S$

(1) **initialize** $J^{(0)}(s) = 0$, $s \in S$, and set $k = 0$;
(2) **for each** $s \in S$, **compute**

$$J^{(k+1)} = \max_{a \in A(s)} r(s,a) + E\left[J^{(k)}(\tilde{s})\right]; \tag{4.11}$$

(3) **repeat (2) until convergence is achieved;**
(4) **for each** $s \in S$ **compute**

$$\mu(s) = \arg \max_{a \in A(s)} r(s,a) + E\left[J^{(k)}(\tilde{s})\right]. \tag{4.12}$$

where $J^*(s)$ is the maximum spectrum utilization with respect to $s \in S$; the state $s = \left\{B^{(1)}, \ldots, B^{(N)}, b\right\}$ is characterized by both the available spectrum fragments $1-N$ and the current requested bandwidth b; S is the set of all possible state and the number of states is finite; $J^{(n)}(s)$ is the optimization result at nth iteration; $r(s,a)$ denotes the throughput with respect to state s and assignment a. $E[*]$ denotes the expectation. The detailed derivation of the algorithm can be found in Appendix D.

4.2.2.1 Numerical Results

Some numerical results are provided to show the performance of the dynamic spectrum assignment solution. At the beginning, the spectrum fragments available to the SM is given as $\left\{B_0^{(1)}, \ldots, B_0^{(N)}\right\} = \{7, 8, 9, 16\}$ MHz. The set of possible requested bandwidth from WSDs is given by $\beta = \{2, 3, 5\}$ MHz with the corresponding arrival probability given by $[0.1, 0.5, 0.4]$.

First, we compute the optimal spectrum assignment policy using the value iteration method in Algorithm 4.1. We plot the value functions for the initial state $s_0 = \left\{B_0^{(1)}, \cdots, B_0^{(N)}, b\right\}$, $b \in \beta$, computed at different iterations in Figure 4.17. The convergence of the algorithm can be observed after 15 iterations.

Next, we compare the optimal spectrum assignment policy with two heuristic policies. The first heuristic policy is to randomly assign the arrived bandwidth request to a spectrum fragment that can accommodate it. It is called "random policy." The second one is to assign the arrived bandwidth request to the smallest fragment that can accommodate it. It is hence termed as "smallest fragment policy." 10 000 computer simulations are conducted and in each simulation the arrival process is generated according to the considered probability distribution. The optimal policy as well as two heuristic ones is executed.

The histograms of the achieved throughputs at the end of different simulations for the three policies are shown in Figures 4.18–4.20, respectively. Note that the maximum throughput that can be achieved by the policies is bounded by the sum

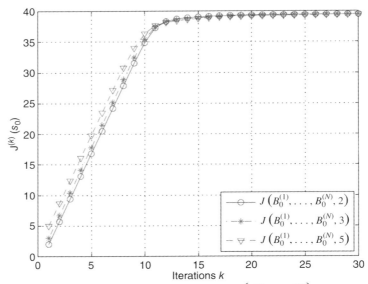

Figure 4.17   The value functions for the initial state $s_0 = s_0 = \left\{ B_0^{(1)}, \ldots, B_0^{(N)} \right\}$, $b \in \beta$ computed at different iterations.

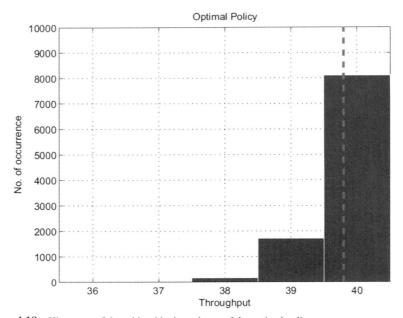

Figure 4.18   Histogram of the achievable throughputs of the optimal policy.

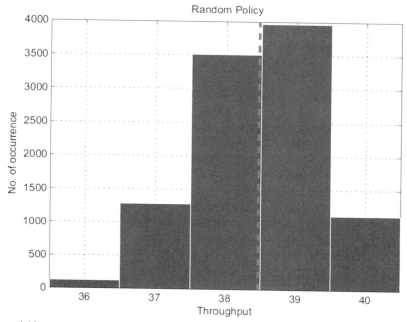

Figure 4.19 Histogram of the achievable throughputs of the random policy.

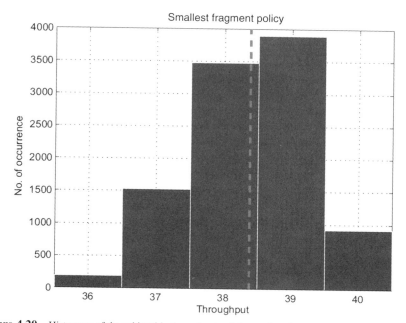

Figure 4.20 Histogram of the achievable throughputs of the smallest fragment policy.

of the bandwidth in the initial fragments, which is 40 MHz in our settings. It can be observed from Figure 4.18 that by considering the distribution of the future arrivals, the optimal policy can better utilize the available spectrum fragments, and it can achieve the maximum throughout of 40 MHz for 81% simulations. In contrast, the random and smallest fragment policies can only fully utilize the available spectrum fragments for 11 and 9% of the simulations, respectively. In addition, we also plot the spectrum utilization of the three policies averaged over 10 000 simulations. They are shown as the dashed lines in Figures 4.18–4.20. We can see that the optimal, random, and the smallest fragment policies can achieve an average throughput of 39.8, 38.5, and 38.4 MHz, respectively. Thus, the optimal policy for the dynamic spectrum assignment achieves the highest spectrum utilization overall.

4.3 NETWORK LAYER

Application layer	REM through WSD networks
	Enhanced WSDB (area-based, embedded broadcast, WSDB-Q)
Network layer	*Robust TVWS networks through advanced routing*
Medium access layer	Dynamic spectrum assignment
	SU networks coexistence based on IPM
Physical layer	Positioning
	Out-of-band leakage control
	Channel aggregation
	Spectrum identification
	TVWS antenna

TVWS provides plenty of spectrum opportunities, and thus is able to support high-throughput wireless applications. However, one of the challenges in the network layer is that the availability of TVWS varies from time to time even in the same geographic location. Therefore, when we build a large-scale network in TVWS, we need to take into account this sort of variation.

In general, the issue of the TVWS network mentioned above could be solved in two kinds of solutions:

1. *Self-Healing-Based Solution.* When the communications are blocked due to the change of availability of operating frequency of a TVWS network, it automatically looks for other channels through reconfiguring the network parameters. With self-healing function, the communication network can be recovered. However, data loss and service suspension are inevitable because the healing action is after the disconnection of the communication link.

2. *Prediction-Based Solution.* Although the availability of spectrum is time-varying, it can be predicted to a certain degree. The simplest case is that PUs' activities are usually informed by themselves or regulators in advance. Even if these activities are not notified, they could be predicted

according to historical records. Predicting PUs activities can help SU networks to establish the alternative route before communication links are gone.

To extend the coverage, the TVWS CRN could be designed as a mesh network. This requirement stimulates the development of CRN routing protocols. The routing protocols in CRN usually combine multiple metrics to determine the route for a transmission [33–40]. The commonly used metrics include delay, hop counts, power consumption, signal-to-noise ratio, location, spectrum availability, and routing stability. Among these metrics, routing stability and spectrum availability are the most CRN-attributed metrics since they are used to reflect the stability of a route based on PU activities. An unstable route will lead to frequent triggering of route reestablishment that consumes network resources and degrades the system performance. Therefore, the straightforward strategy to achieve routing stability is simply to avoid choosing the route in which PU activities are often observed.

The spectrum-tree-based on-demand routing protocol (STOD-RP) [36] uses a routing metric that combines route stability and end-to-end delay. For the stability part of the routing metric, the stability in terms of link's available time, denoted as T, is used as a denominator in the routing metric. Since T is defined as the time duration during which a spectrum band is available to the link that can be predicted from the statistical history of PU's activities, it can become infinity for a link having no PU activity at all. As a result, the cost of the link will become zero even if the link itself is very lossy. The work proposed by Filippini et al. [41] introduces the new cost that will be paid if a route needs to be changed. The link cost is defined as $(C_{Sw} + \alpha C_{Rep})/E[TTS]$, where C_{Sw} is the cost of switching from the current link to another links, C_{Rep} is the expected cost to repair the link in the future, and α allows gauging of different cost contributions. Since $E[TTS]$, the average time the link remains stable, is used as a denominator in the metric too, this protocol also has the same problem of overconsidering the PU activity as STOD-RP does.

In order to enhance the service quality experienced by the CRN user, a novel CRN routing protocol named "dynamic path switching (DPS)" is designed [37] to utilize the good-quality link as much as possible even if there is a lot of PU activities observed on the link. To alleviate the cost induced by the PU interruption, the DPS routing protocol switches aggressively among available paths. This strategy is motivated by the fact that the cost for the proactive path switching is always less than the cost for the passive path switching, where the path is forced to switch due to the link unavailability.

There are three core strategies applied to the DPS routing protocol: they are (1) predicting the stable time for the link and path; (2) utilizing the good-quality link as much as possible; and (3) aggressively switching among paths, since the path switching cost is always smaller than the service suspension and route repair costs induced by the occurrence of PUs.

To evaluate the performance of the DPS routing protocol, several experiments are conducted to compare the performance of the DPS routing protocol (with the t_{thres} value equal to 5 s) with the performance of other routing protocols, including two CR routing protocols and one traditional routing protocol. The traditional routing protocol used in the experiments is the high-profile AODV routing protocol that considers both the hop-count and link quality as its routing metrics. The link quality of a link is estimated by counting the number of lost packets on the link in a given time.

As to the first comparative CR routing protocol, it is built based on the STOD-PR [36] protocol, which uses the "spectrum available time" as the denominator when calculating the link-cost. Therefore, it can be used to represent the typical type of stability routing protocol that considers the PU influence and the stability characteristic as the denominator in the link-level routing metric (e.g., the work proposed by Filippini et al. [41]). In the following description, the first comparative CR routing protocol is named as "CR Stability Routing (strong ver.)" instead.

The second comparative CR Routing Protocol is constructed based on the Coolest-Path [42] protocol, with the "minimum highest spectrum temperature path" configuration applied to the protocol. Compared to the "CR Stability Routing (strong ver.)" Protocol, this type of CR routing gives a relative light weight to the PU influence when constructing the link-level routing metric. In the following description, this kind of CR routing protocol is named as "CR Stability Routing (mild ver.)" instead.

As shown in Figure 4.21, a real Wi-Fi network is constructed for a series of experiments that will be described later. In the network, six Wi-Fi routers are deployed with the ad-hoc mode configuration. Two of the routers (A and B) are wired-connected to the centre aggregator, while the others are connected to each other with wireless connections. All routers are configured to operate in the IEEE 802.11g mode with the operating frequency equal to 2.427 GHz (channel 4), with the operating bandwidths equal to 20 MHz, and with no request to send/clear to send (RTS/CTS) exchange. For simplicity, the whole network is constructed as a single-channel network. This means when a PU activity is presented in a certain link, that link will become unavailable until the PU becomes inactive again.

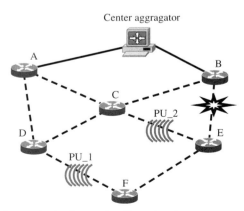

Figure 4.21   The environment setup for a series of experiments.

To simulate PU activities, two jamming sources are applied to the link between the router D and F, and to the link between the router C and E. The PU activity simulated between D and F is denoted as PU_1, and the PU activity simulated between E and F is denoted as PU_2. The activity density of PU_1 is always twice the density of PU_2, and each PU activity lasts constantly 25 s in all experiments. Moreover, the link between B and E is configured as an error-prone link by introducing a noisy source that keeps the packet drop rate on the link equal to 0.5.

To observe the throughput results, a big file is uploading from the FTP client located at the router F to the FTP server located at the centre aggregator PC. A PING testing is also conducted on the router F and the centre aggregator PC to obtain the results of the packet loss rate and averaged packet round-trip delay. By adjusting the PU activity density of PU_1 and PU_2 accordingly, the results of accumulated throughput, packet lose rate, and averaged round-trip delay within the duration of 15 min are observed as follows.

The FTP throughput results for different routing protocols are shown in Figure 4.22. In the figure, the notation (x_1, y_1) in the x-axis denotes that the arrivals of PU_1 follow the Poisson process with the rate equal to $1/x_1$ per second, and the arrivals of PU_2 follow the Poisson process with the rate equal to $1/y_1$ per second. This means the left-hand side on the x-axis represents a lower PU activity density, and the right-hand side represents a higher PU activity density. The results in Figure 4.26 show that the DPS routing protocol and the traditional routing protocol

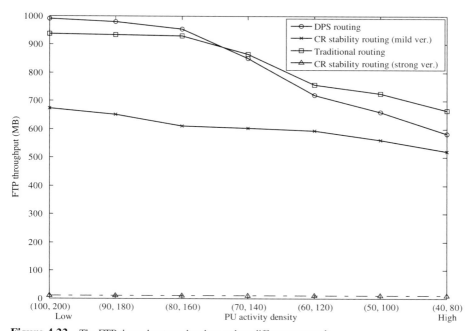

Figure 4.22   The FTP throughput results observed on different protocols.

have similar throughput performance and they outperform other CR Stability Routing Protocols. The DPS protocol performs a little better in low PU activity density situations, while the traditional protocol performs a little better in high PU activity density situations. This is because when PU activity density becomes high, the DPS will aggressively switch from the optimized path to other suboptimized paths, which lowers the utilization rate of the optimized path more than the traditional routing protocol does. The CR Stability Routing (strong ver.) obtains very poor results because the transmission always picks the path that contains the error-prone link (the link between B and E). The error-prone link causes a huge number of lost packets and retransmissions that will in turn degrade significantly the FTP performance.

The packet loss rate (of PING) results are illustrated in Figure 4.23. In either low or medium PU activity density conditions, the DPS routing protocol has the smallest packet loss rate among all routing protocols. However, in high PU activity density conditions, the CR Stability Routing (mild vers.) protocol has the smallest loss rate which is slightly better than that in the DPS routing protocol. It is because in such conditions, the DPS protocol will have a higher chance to pick the error-prone link to construct the path (when it predicts the occurrence of PU_1 and PU_2 will overlap), while the CR Stability Routing (mild vers.) Protocol will pick the error-prone link only when both PU_1 and PU_2 actually occur at the same time. The CR Stability Routing (strong ver.) Protocol again has the worst packet loss rate since it always picks the path containing the error-prone link.

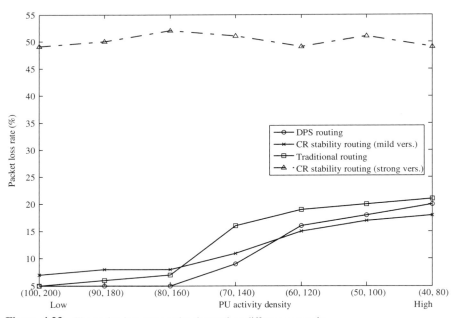

Figure 4.23   The packet loss rate results observed on different protocols.

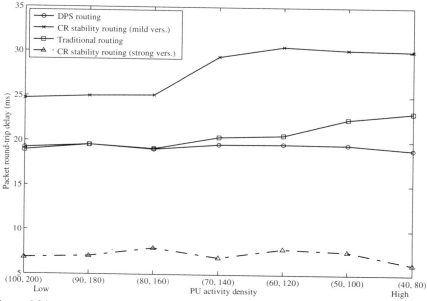

Figure 4.24 The packet round-trip delay results observed on different protocols.

The last results shown in Figure 4.24 are the packet round-trip delay (of PING) comparisons. We can see that the DPS protocol is very resistant to the PU activity change since it almost maintains the same averaged delay around 19 ms. The CR Stability Routing (mild ver.) protocol has the largest averaged delay because it often picks the path containing the link between the router C and E, and this path obviously requires more transmission hops than other paths do. Although the averaged delay seems to be excellent in the CR Stability Routing (strong ver.) protocol, about 50% of the packets are lost and not counted into the delay; therefore, it is not meaningful to interpret the results in that way.

4.4 APPLICATION LAYER

As mentioned previously, since current spectrum sensing technology faces the challenge of reliability, it is required by regulators that all WSDs except very small power ones shall periodically acquire information from the WSDB before using TVWS channels. This becomes a feature of TVWS applications. In the physical layer, we introduced WSDB as one of the approaches to obtain spectrum information. In this section, we will present some enhanced functions of WSDB in the application layer.

The interface between WSDB and WSD is shown in Figure 4.25. A WSD could access one or multiple WSDBs though a WSDB/WSD interface. There are a few existing interfaces such as GDD adopted by IEEE 802.11af and PAWS proposed by Internet Engineering Task Force [43] as introduced in Chapter 3.

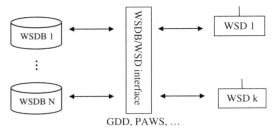

Figure 4.25   Illustration of relationship between WSDB and WSD.

The interfaces between WSDB and WSD have not been unified yet. Also, the information exchanged between WSDB and WSD currently is processed in the application layer.

In the next subsections, we will introduce the enhanced WSDBs and radio environment map (REM) based on WSDB.

4.4.1   Enhanced WSDB

Application layer	REM through WSD networks
	Enhanced WSDB (area-based, embedded broadcast, WSDB with QoS)
Network layer	Robust TVWS networks through advanced routing
Medium access layer	Dynamic spectrum assignment
	SU networks coexistence based on IPM
Physical layer	Positioning
	Out-of-band leakage control
	Channel aggregation
	Spectrum identification
	TVWS antenna

4.4.1.1 Area-Based WSDB

A TVWS network typically includes many devices (fixed or mobile) operating in an area. For fixed devices, the locations are known and unchanged within the area. For mobile or portable devices, the area might not be easily confined, but some applications do have prior knowledge about the path of the mobile devices, for example, devices mounted on buses that operate in a campus. For cases where the areas of operation of a TVWS network are known, it is possible to perform WSDB inquiry in an area rather than device by device. Doing so not only reduces the number of access to the geolocation database (WSDB) but also reduces the number of devices requiring internet access to WSDB. We are introducing the methods for the

WSDB to determine a list of available channels for secondary TVWS or other dynamic spectrum access networks that require operation within an arbitrary area.

In the FCC report [44], it is stated that a mobile SU device could obtain spectrum information for an area from the WSDB. The FCC proposed WSDB is different from the FCC. In the FCC proposal, a geographic area is divided into pixels (100 m × 100 m) and every pixel will contain a maximum SU emission level for all TV channels. When a SU queries the WSDB, the WSDB will reply back the SU with the maximum power it can transmit in the various TV channels based on which pixel the SU is located in. Hence, for an area, the WSDB can determine the maximum power of the SUs in the various TV channels based on the information of the pixels that are inside the area. A figure taken from the FCC report is shown in Figure 4.26. This method is applicable for countries, for example in Europe, having such a map with pixels information when planning for the digital terrestrial television coverage.

In the FCC report [45], it is stated that it is possible for a SU to request for available channels for an area from the WSDB to support the operation of mobile SUs. The area-based WSDB can determine the available channels for an arbitrary area defined by the SU. There are basically two ways to define the area:

1. The area is defined by a center point and a radius, which are provided by the SU to the WSDB. The WSDB uses the two parameters to determine whether the channels are available for the SUs to operate within the area

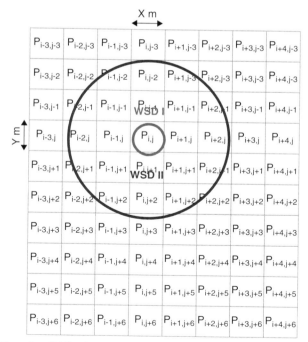

Figure 4.26 Channel availability for an area proposed in FCC report 159 where WSDs are SU devices.

of circle formed by the two parameters. This can be a quick solution with a low computation complexity if a suitable subset of "keep-out contour coordinate points" (KOCCPs) is selected instead of complete searching.

2. The area is defined by a series of coordinate points that bound an arbitrary area. The SUs provide these coordinate points to the WSDB to request for available channels. By combining the information of the channel availabilities of each of these points, the WSDB then calculate which channels are available for the whole area. Depending on the density of the coordinate points and the shape of the area, the computation complexity varies and uncertainty may happen. Thus, it needs an additional check to determine any KOCCP that falls inside the area to further ensure that there is no overlap between the defined area and the PU keep-out contour.

The detailed calculations for the area-based WSDB can be found in Appendix E.

4.4.1.2 Broadcast WSDB

As we know, WSDs except for small power ones need to access WSDB periodically for acquiring the latest TVWS information. However, one of the fundamental drawbacks of current WSDB mechanism is that it requires the WSDs to access the database through Internet. This causes some issues:

- An Internet link between WSDB and the fixed or mode II WSD is needed.[3] This raises a question. If a user has already Internet access capability, why does it need the TVWS communications? This reduces the user's interest in the TVWS.
- With the connection to the Internet, there is a risk of backdoor hacking and thus increases the risk of leakage of user confidential information.
- Certain deployment may not have Internet link readily available.

All the above discourages users who require higher level of security to use TVWS.

A possible way to solve the above problem is to broadcast the primary users' information through beacon signals. The beacon signal can include the channel occupancy information, transmission power, locations, and coverage area of the primary transmitters, and so on. The WSD can obtain the available TVWS information through one-way broadcasting instead of Internet access.

First, let us look at the existing beacon approaches. Some types of beacon transmission are discussed in literature [10,11,46–48]. They can be classified into three categories:

(1) Each licensed primary transmitter broadcasts the beacon signal that contains the transmitter's own information [10,11,46,47]. In this case, since

---

[3] Note that mode I WSD does not require WSDB access and thus a single Internet link typically suffice in a TVWS cell.

the beacon signal of each transmitter only carries its own information, no connection to WSDB is required. A WSD can access the PU's channel when PU's beacon is not heard by the WSD.

(2) A beacon transmitter broadcasts the list of channels that can be used by any WSD within the beacon transmitter's coverage area [48]. In this case, the beacon transmitter needs to be connected to a WSDB to obtain the list of available channels within its coverage area.

(3) A beacon transmitter broadcasts all PU transmitters' information (frequency band, location, and coverage area of each licensed transmitter) within its coverage area using dedicated channel [11]. A WSD needs to use its own location information together with the received beacon information to determine whether a channel is accessible. In this case, the WSD may need to compute itself if it is within the coverage region of certain licensed transmitter. In the approach, the geolocation information is used. The WSD must obtain the WSDB information before it accesses TVWS. Thus, it eliminates the risk of collision. However, this requires dedicated beacon transmitter that occupies an additional frequency resource.

In fact, the WSDB can be embedded into multiple public broadcasters' existing signals to broadcast WSDB information without the need for dedicated channel for WSDB broadcasting. The difference between this approach and the tradition primary beacon approach is that there is no change to the PUs. The broadcaster is not seen as the PU but serve as a WSDB in another form. This is called embedded broadcast WSDB.

The WSDB can be embedded in an analogue TV signal through teletext function, in a Digital Video Broadcasting (DVB) signal through Program-Specific Information/Service Information (PSI/SI) function, in a digital audio broadcast (DAB) signal through Fast Information Channel (FIC)/Programme-Associated Data (PAD), or in HD radio through L4 encoding. Some details are provided in Appendix F.

Figure 4.27 shows the concept of broadcast WSDB system. The WSDB is connected to some or all public broadcasting stations. The WSDB information including the available channels and each location areas are broadcasted through these stations by embedding the WSDB data into the existing broadcasting channels. The WSDs can obtain the available channel information through receiving the public broadcasting and decode the WSDB data and select the suitable channel for TVWS communications.

The broadcasted information contents for embedded broadcast WSDB consist of

(1) Transmitter identification (ID) with location information, where the transmitter is the broadcaster broadcasting the WSDB.

(2) TVWS WSDB data. Here the TVWS WSDB data have three possible formats:

   (a) *Pixel-Based*. Namely, the whole area serviced by the WSDB is divided into a number of pixel areas (similar to the case of WSDB

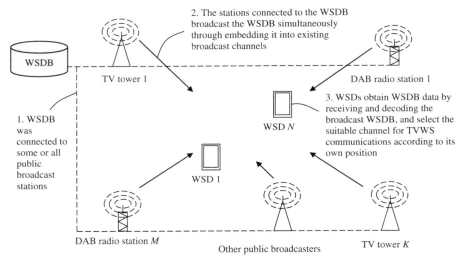

Figure 4.27   Concept of the broadcasting TVWS system through embedding it into existing broadcast channels.

from Ofcom), and the WSDB provides the information on the location of each pixel area and the available channels with power limitations in each pixel area. Figure 4.28 illustrates service area segmentation in the pixel-based WSDB. The pixel area can take different shapes. Two example shapes are shown here. The location of each pixel is determined by its corner locations, which are in details P1–P4 for square pixel version or P1–P6 for cellular pixel version. The corner points can be given directly or through central location of the pixel plus the radius and the pose information.

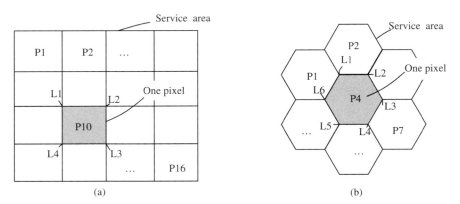

Figure 4.28   Illustration of service area segmentation in the pixel-based TVWS WSDB.

(b) *Contour-Based.* Namely, the WSDB consists of primary users' protection contours in the service area. WSD will rely on this information to determine the available channel.

(c) *Transmitter Parameter-Based.* Namely, the WSDB consists of the primary users' emission powers, locations, antenna heights above average terrain, radiation pattern, and so on in the service area. WSD will rely on this information to calculate the contours of the primary users and then determine the suitable channel itself.

To reduce the broadcasting load, a countrywide WSDB can be separated to N service areas, where $N \geq 1$. A WSDB broadcaster broadcasts only the WSDB data for the local service area covered, for example, a large city, areas nearby the broadcaster, or simply the intended coverage area of the particular broadcast station.

The embedded signal contents are suggested in Figure 4.29. The transmitted signal could include some or all of the listed items.

WSDB data can be a simple version, including just PU frequencies, powers, and locations. In this case, WSDs need to calculate themselves the available frequencies at their own locations after they received the broadcast signals. WSDB data can also be an advanced version, including the pixel-based available channels that are calculated by the server. The time reference is optional and is for the WSD to carry out self-positioning, which needs to receive broadcast signals from at least three towers, without additional positioning device such as GPS. After a WSD know its own location and the available channels with respect to locations, it can select the suitable frequency for communications.

The broadcast WSDB signal should be encrypted to protect it from illegal use.

WSD Quality Control in Broadcast TVWS System   Since broadcasting is a one-way signal transmission from WSDB to WSDs, the WSDB is not able to know about WSD information during each spectrum access application. There must be another way to control WSD's quality. This can be done when the WSD registers to the WSDB as a member. A WSD's performances, including the maximum emission power, out-of-band leakage class, receiver's SNR level, and so on, have to be provided. The WSDB will approve the WSD based on a set of criteria agreed with the regulators. Only the approved WSD will be provided a decryption code to recover the WSDB data.

First Adjacent Channel Usage Versus WSD Location   The out-of-band (OOB) interference of a WSD may have serious effects when the WSD works at the first adjacent channels of a primary user or close to the primary user who is having a small receiving power, since it is at the edge of the protection contour.

WSDB data	Time reference data
Encryption	

Figure 4.29   An example signal structure of broadcast WSDB.

However, the OOB interference effect is reduced when the WSD and primary user are in the protection contour and far from contour boundary.

If the WSD is far from the boundary of the protection contour, the first adjacent channels of the protected channel could be used by the WSD with less limitation. Otherwise, the first adjacent channels are prohibited to be used by the WSD, or being used with significant limitations.

4.4.1.3 WSDB with QoS

As mentioned in Chapter 1, current WSDB only protects the primary users. It does not guarantee the performance of the SUs. While this is understood from the current regulation practices, it has to be enhanced to allow differentiated access even for SUs. Otherwise, QoS may become a big issue when TVWS is widely adopted, especially for services that require a certain QoS.

In this section, we discuss a new method called WSDB with QoS (WSDB-Q) that can grant access to higher priority communication via reserved channels based on a set of criteria predefined in order to maintain the QoS of communication systems accessing TVWS spectrum. The set of criteria covers regulatory, technical, and commercial aspects. Singapore IDA's HPC regulation is a type of WSDB-Q.

Figure 4.30 shows an example of the list of the WSDB-Q [49,50]. From the list of original vacant channels, the WSDB-Q administrator, based on certain guideline or regulation, reserves certain channels for higher QoS communication. The number of channels to be reserved could be a regulatory parameter or based on the dynamic needs of certain locations.

When a device queries the WSDB-Q, it can only use vacant channels for communications by default. Whenever the device needs to send data, it should use only the vacant channels returned by the WSDB-Q no matter whether the data are urgent or not. In case all the vacant channels are busy, the device will then request to use the reserved channels based on a set of predefined criteria only if the device has high priority and urgent data to be sent out or received.

The followings are some possible predefined criteria for accessing the reserved channels.

Regulatory Approach   To determine which device is allowed to access the reserved channel, a possible approach is through regulatory and standardization control. A regulatory/standardization/certification body will certify devices based on their class of access and the need for urgent communication. Special class ID

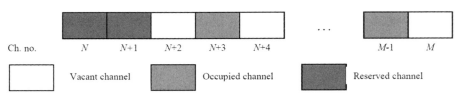

| Ch. no. | N | $N+1$ | $N+2$ | $N+3$ | $N+4$ | ... | M-1 | M |

☐ Vacant channel   ▨ Occupied channel   ▉ Reserved channel

Figure 4.30   Example of list of channels returned by WSDB-Q.

will be issued for devices that meet the criteria to access the reserved channels. A small fee may be imposed to deter nonurgent devices from trying to access these reserved channels. In this case, only these types of devices could access the reserved channels (via message exchanges with WSDB-Q), thus, freeing up nonurgent devices from occupying the reserved channels (part of the original vacant channels) and creating interference to other communications.

Technical Approach   Another approach to control the access to the reserved channels is through technical means. A device that requires urgent communication may send a request to WSDB-Q to access the reserved channel either in real time or at a prenegotiated time. The WSDB-Q may grant access to the reserved channels after certain technical criteria are met, for example, the WSDB-Q receives many "feedback" that certain channels are busy, or the device has retried the channels for n times, etc. The type of communication and accesses may be recorded for postprocessing and further improvements whenever required.

Commercial Approach   The third approach is a market-driven approach. A small fee (the amount may be adjustable based on demand) is imposed whenever a device accesses a reserved channel. For devices that are really in need to access the channel for urgent communication, this small fee may be acceptable. In this case, we could deter unnecessary access to the reserved channels even when the vacant channels are still usable. The mechanism on the fee settlement is not part of the discussion of this book and is left to implementation.

Hybrid Approach   A hybrid approach combining two or more of the above approaches could also be used to ensure better spectrum utilization while maintaining QoS via the WSDB-Q.

On top of the above, the WSDB-Q administrators are also allowed to change the number of reserved channels under certain guideline provided by the regulators. Below are some possible approaches.

Fixed Allocation   The regulator fixed certain number of channels as reserved channels. The WSDB-Q cannot change the number of channels at its own wish. Whenever required, the WSDB-Q has to request for changing the number of reserve channels from the regulator based on demand in offline manner.

Semidynamic Allocation   The WSDB-Q is allowed to change the number of reserved channels based on a range of reserved channels provided by the regulator. In this case, the WSDB-Q will determine the number of reserved channels based on real-time demand from the devices. The WSDB-Q shall reserve the least amount of reserved channels whenever possible.

Fully Dynamic Allocation   The WSDB-Q is allowed to change the number of reserved channels up to the maximum channels available. Similar to the above, the

WSDB-Q will determine the number of reserved channels based on real-time demand from the devices under the guidance of a set of rules set by the regulation.

WSDB-Q also addressed a major concern from the industries on spectrum certainty, that is, at least a TVWS channel is available for WSDs in time or geographic locations as some applications may require more certainty in transmission. This might be a concern to other WSD users in other countries as well although most existing TVWS regulations worldwide have not addressed this concern yet.

The Singapore regulator, IDA, is the first regulator to adopt WSDB-Q approach where they proposed a new type of TVWS channels called the high priority channels (HPCs) that are always available to the WSDs. WSDB providers may impose fees to WSDs for using the HPCs. For a start, IDA has proposed to designate two HPCs within the TVWS channels for WSDs that are willing to pay for higher level of spectrum availability and access.

The fundamental task of the WSDB is to provide WSDs the information of available TVWS channels at their locations. With the introduction of HPCs, it created a new business opportunity for WSDB provider to earn revenue from WSDs that do not mind paying a fee for better QoS. In the future, pay-per-use unlicensed channels may also become a common feature for spectrum sharing if the demand for better QoS in unlicensed channels becomes high. In view of this development, how WSDB provider could maximize its revenue with the management of HPCs access was also investigated.

A hybrid pricing scheme for WSDB provider was proposed in [51], where the WSDB provider will pay a fixed fee to regulator to reserve TVWS channels and then sublet the reserved channels back to WSDs for a fee. Each WSD in these reserved channels can occupy parts of the TV channels by itself without sharing with other WSDs. The rest of WSDs can use the nonreserved TVWS channels for a fee depending on the number of queries. A two-dimensional exhaustive search is used to find the optimal fee for WSDs and the amount of TVWS channels that the WSDB provider needs to reserve. The operating model in [51] is different from HPCs and the criterion for WSDs to justify their rewards for a fee is different from HPCs where they are based on channel capacity, while HPCs consider probability of successful packet transmission.

In this section, we introduce the system model of HPCs, their fee and reward function, and further provide the mathematical formulation of maximizing the revenue of WSDB provider and also the solution to the problem followed by numerical results of the optimal fee and revenue.

We consider that the WSDs coexist within TVWS channels via carrier sense multiple access with collision avoidance (CSMA/CA) protocol. Each WSD will transmit a packet in a slot time with a probability τ that is a function of the contention window in the CSMA/CA protocol [51]. In a constant backoff window problem, it was found that $\tau = \frac{2}{W+1}$, where W is the minimum contention window [51]. Therefore, the average probability of a WSD transmitting a packet successfully in normal TVWS channels is given as

$$P_N = (1 - \tau)^{(n_N/m_N)-1}, \tag{4.13}$$

where n_N is the total number of WSDs in normal TVWS channels and m_N is the total number of normal TVWS channels. Similarly, the average probability of a WSD transmitting a packet successfully in HPCs is given as

$$P_H = (1 - \tau)^{(n_H/m_H)-1},$$

(4.14)

where n_H is the total number of WSDs in HPCs and m_H is the total number of HPCs. Let the ratio of P_H to P_N be the reward function for a WSD to switch from normal TVWS channels to HPCs, which is given as

$$R = (1 - \tau)^{(n_H/m_H)-(n_N/m_N)}.$$

(4.15)

The fee that a WSD is willing to pay WSDB provider for using HPCs will be a function of the reward function R. When $R \leq 1$, there is no benefit in switching from normal channels to HPCs, since the probability of transmitting a packet successfully in normal channels is higher or the same as in HPCs. Hence, WSDs will not pay any fee to move to HPCs when $R \leq 1$. When $R > 1$, we assume that WSDs are willing to pay a fee proportional to R until the fee reaches a maximum limit f_M. This is the price cap of the fee that a WSD is willing to pay regardless of how high R is. In practice, WSDs may listen to the TVWS channels to estimate P_N and P_H and in turn estimate the reward function R.

Other than comparing the reward of HPCs over normal channels, another criterion for a WSD to switch from normal channels to HPCs is that the P_H must be greater than or equal to a given threshold $\overline{P}_H$. This criterion is to justify the scenario where P_H may be too low to achieve adequate QoS required by WSDs but still achieve a high R simply because P_N is very low. Hence, although the probability of transmitting a packet successfully in HPCs is higher than normal channels, WSDs will still not pay any fee to use HPCs if their required QoS is not met in HPCs. The fee that WSDs are willing to pay based on the above criteria is given as

$$f = \begin{cases} f_M & \text{if } R \geq R_T, P_H \geq \overline{P}_H \\ \dfrac{f_M}{R_T - 1}(R - 1) & \text{if } 1 < R < R_T, \ P_H \geq \overline{P}_H \ . \\ 0 & \text{otherwise} \end{cases}$$

(4.16)

where R_T is the reward value corresponds to the point where the fee first reach its maximum limit f_M.

For a WSDB provider, the objective is to set a fee that will maximize its revenue based on the number of WSDs. We have a detailed derivation on revenue maximization based in the system model in Appendix G. Here, some numerical results are presented to show how the WSDB provider can win through the HPC.

Let us set the number of normal TVWS channels and HPCs to be 3 and 2, respectively. The fee is normalized such that $f_M = 1$ and R_T is set to be 10 such that if P_H is more than 10 times better than P_N, the WSDs will not increase their fee above f_M to use the HPCs. The minimum required probability to transmit a packet successfully in HPCs is set at $\overline{P}_H = 0.6$.

The optimal revenues achieved under different values of τ are shown in Figure 4.31. The optimal revenue results obtained using exhaustive search are also plotted as comparison and it is shown the derived policy matches the optimal solutions from exhaustive search. It is observed that there is a ceiling to the revenue for every τ value. This is because of the QoS constraint that limits the number of WSDs in HPCs and this number multiply by f_M is the revenue ceiling. The revenue ceiling is higher for smaller τ since the probability of each WSD transmitting a packet is lower, it allows more WSDs to operate in HPCs with the given QoS constraint. It is observed that when the total number of WSDs is small, it is better for the WSDB provider that the WSDs are transmitting packets at a higher frequency since the normal TVWS channels will get congested more easily and WSDs are more willing to pay to move to HPCs. However, when the total number of WSDs is large, the reverse is true because of the QoS constraint.

The optimal normalized fees obtained by the derived policy and the exhaustive search with different values of τ are shown in Figure 4.31. Similarly, it is observed that the derived policy is the optimal solution. From the figure, it is shown that the WSDB provider in general should increase the fee as the number of WSDs increases. However, there are some exceptions, for example, when $\tau = 0.2$ and number of WSDs is 35, the optimal normalized fee is 0.6 but when the number of WSDs is 36, the optimal normalized fee falls to 0.53. This is because by lowering the fee, it will attract more WSDs to join HPCs and this will increase the overall

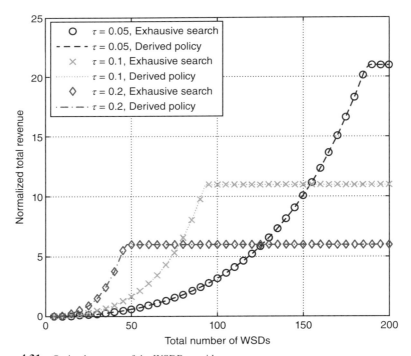

Figure 4.31   Optimal revenue of the WSDB provider.

revenue even though the fee from individual WSD is lower. In the above scenario, when the number of WSDs is 35 and with a fee of 0.6, it attracts 4 WSDs to use the HPCs but when the number of WSDs is 36 and with a fee of 0.53, it will attract one more WSDs and hence the total revenue actually increases. When τ is high, WSDB provider should set a higher fee because the probability of successful packet transmission decreases at a faster rate when the number of WSDs increases and therefore, WSDs are more willing to pay higher fee to switch to HPCs to improve their probability of transmitting a packet successfully.

The number of WSDs in HPCs based on the optimal fee provided in Figure 4.32 is shown in Figure 4.33. It is observed that the number of WSDs in HPCs increases as the total number of WSDs increases although the fee is increasing as well. The number of WSDs in HPCs increases until it reaches the maximum number defined by the QoS constraint.

HPCs are new features proposed by IDA that offer WSDB providers a new business opportunity to increase revenue. The WSDB provider's revenue can be maximized by a well-designed policy. It is shown that WSDB providers should not always increase the fee when the number of WSDs increases as reducing the fee to attract more WSDs to use HPCs may actually increase the revenue. It is also shown that when the total number of WSDs is low, it is better for the WSDB provider if the WSDs are transmitting more frequently. However, when the total number of WSDs is high, it becomes worse.

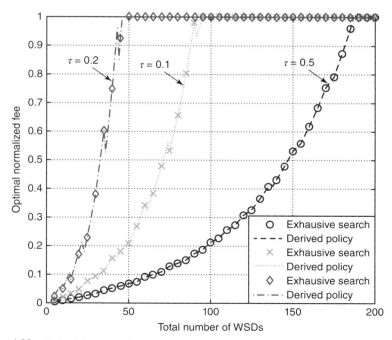

Figure 4.32 Optimal fee set by the WSDB provider.

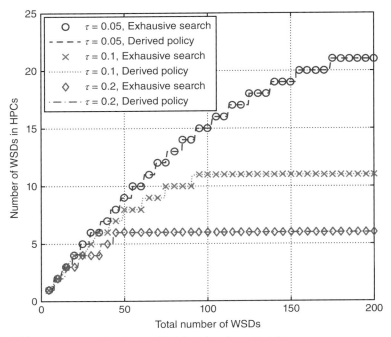

Figure 4.33 Number of WSDs that use HPCs based on the optimal fee.

4.4.2 REM Through WSD Networks

Application layer	REM through WSD networks
	Enhanced WSDB (area-based, embedded broadcast, WSDB-Q)
Network layer	Robust TVWS networks through advanced routing
Medium access layer	Dynamic spectrum assignment
	SU networks coexistence based on IPM
Physical layer	Positioning
	Out-of-band leakage control
	Channel aggregation
	Spectrum identification
	TVWS antenna

Besides providing WSDs with a list of available channels for communication, WSDB also has another key value to be exploited, that is, the big data that it collects. When the number of WSDs connected to WSDB is large and the WSDs feedback certain information, for example, channel quality, to the WSDB, the WSDB actually becomes a virtual sensor network, that is, a sensor network without the need to maintain physical sensors. The information collected by the WSDB forms the radio environment map (REM) that can be exploited for applications beyond TVWS. As examples, we present two new potential applications below.

4.4.2.1 WSDB-Based REM for Other Frequencies

A straightforward application is to use the feedback of WSDs on radio condition to other wireless networks in different frequencies for them to carry out cell planning, to establish RF fingerprint, and so on. For example, we can use a TVWS network to support REM for a LTE network. This requires path loss profile translation from TVWS to other frequency bands. The relevant radio measurements collected via TVWS WSDB are first used to estimate the path loss profile at the TV frequencies. A general path loss profile is given as

$$PL = 10\alpha \log d + C, \tag{4.17}$$

where d is the distance between the transmitter and the receiver, α is the path loss exponent, and C is a constant fixed loss. The distance d can be computed from the location information of the WSD transmitters and receivers that are stored in the WSDB. The path loss exponent is commonly estimated using least-squares linear regression fit to the radio measurements [52,53]. The fixed constant loss C can be modeled by

$$20 \log(4\pi d_0/\lambda), \tag{4.18}$$

where λ is the wavelength and d_0 is the close-in distance [52]. The fixed constant loss C can also be jointly estimated with α using least-squares linear regression fit [53,54].

After estimating the path loss profile of the TVWS network, it will be used to generate the path loss profile of the LTE BS. There are existing methods to convert a path loss profile from one frequency to another given the same geographic area. One common method is to use the equation given as

$$PL_{\hat{f}} = PL_{\check{f}} - \beta_1 \log \check{f} + \beta_2 \log \hat{f} + \varepsilon, \tag{4.19}$$

where $PL_{\hat{f}}$ is the path loss profile at frequency $\hat{f}$ to be estimated, $PL_{\check{f}}$ is the path loss profile at frequency $\check{f}$ that is known, and β_1, β_2, and ε are constant values depending on the propagation models used. For example, to convert any path loss profile between frequencies 150 and 1500 MHz based on Hata propagation model, the parameters are $\beta_1 = \beta_2 = 26.16$ and $\varepsilon = 0$. To convert any path loss profile from between 150 and 1500 MHz to 1500 and 2000 MHz based on Hata and COST-231 propagation models, the parameters are $\beta_1 = 26.16$, $\beta = 33.9$, and $\varepsilon = -23.25$. There are other methods to convert a path loss profile from one frequency to another such as using the median attenuation relative to free-space values in Okumura model [54] or the equation provided in the paper [55]. With the estimated path loss profile, the LTE BS is able to perform REM based on its location and transmit power.

Another possible method in using the relevant radio measurements to construct REM in other frequency is to first convert the individual collected radio measurements to radio measurements of another frequency. For example, using RSS measurements that are collected through the WSDB-based network can be used to estimate the RSS of another frequency using the same approach as in Equation 4.19 given as

$$RSS_{\hat{f}} = RSS_{\check{f}} - \beta_1 \log \check{f} + \beta_2 \log \hat{f} + \varepsilon, \tag{4.20}$$

where $RSS_{\hat{f}}$ is the received signal strength at frequency $\hat{f}$ to be estimated and $RSS_{\check{f}}$ is the received signal strength at frequency $\check{f}$ that is gathered using WSDB-based networks. The estimated $RSS_{\hat{f}}$ can then be used in the REM of another network coexisting in the same geographic area.

Cell Planning   An important information in cell planning for any system is the signal coverage of the base station at a given location. However, to know the signal coverage of the base station at a given location requires massive on-site data gathering around the base station. This is a manpower- and time-consuming task. Even after determining the signal coverage of the base station with the massive on-site data gathering, the signal coverage of the base station may change when the landscape around it changes. This is especially valid in urban city where there are always new buildings being constructed.

The amount of manpower and time can be reduced by using the WSDB system to gather the data instead. The WSDB system identifies the WSDs that are located around the base station and requests the WSDs to collect the radio measurements. Although the radio measurements are from a different frequency, they can still be used to predict the path loss profile of the base station based on our technology. Once the path loss profile of the base station at its location is known, its signal coverage can be estimated from the path loss profile and its transmit power.

Establish the RF Fingerprints   RF fingerprint refers to a method for localization. It consists of two steps: first, the network characteristics, such as the RSS values to different transmitters at different locations of the network, have to be premeasured and stored in the database; then, a device's location is estimated by comparing the current measurement with the prestored measurements in the database [56]. RF fingerprint has been used primarily for indoor positioning [57] and places where not enough satellites are visible due to the blockage by high-rise buildings in urban areas [58].

Conventionally, the RSS values required for fingerprint are obtained from a one-time site survey and regular updates subsequently. The TVWS REM solution allows the RSS values to be constantly updated and adapt to the changes in the physical environment. It is possible to estimate RSS values and establishing the RF fingerprints for multiple systems operating over the same geographic area but using a different frequency.

For simple illustration, let us denote the WSDB network as Network A and the network colocated with Network A that requires REM as Network B. The procedure for constructing REM using Network A for Localization in Network B is as follows:

(1) Device n of Network A at location (x^n, y^n) measures RSS to M base station in Network A $\left(p_1^n, p_2^n, \ldots, p_M^n\right)$ and reports them to Network A REM.

(2) REM stores the information and converts the RSS values in Network A to the RSS in Network B $(\tilde{p}_1^n, \tilde{p}_2^n, \ldots, \tilde{p}_M^n)$.

(3) Device at an unknown location in network B measures its RSS to M base station in network B $\left(q_1, q_2, \ldots, q_M\right)$ and reports to REM.

(4) REM computes the device location by comparing the current measurement $\left(q_1, q_2, \ldots, q_M\right)$ with the converted RSS values in REM $\{(x^n, y^n, \tilde{p}_1^n, \tilde{p}_2^n, \ldots, \tilde{p}_M^n)\}$.

4.4.2.2 Coverage Hole Detection for Cellular Systems

Coverage estimation is one of the key problems in deployment and operation of cellular networks. Although sophisticated planning tools incorporating building and terrain aware propagation models are often used by the operators to estimate the signal coverage, coverage holes are nevertheless occasionally emerged. These holes can result from changes in the propagation environment or other causes. To discover these coverage holes, drive tests are used as the routine part of network diagnostics, or when customer complaints are received. However, using the drive test as routines is always expensive and time-consuming, and using it as responses to customer complaints still requires the exact location information. AWSDB-assisted coverage hole detection system which can proactively indicate the location of the coverage hole caused by obstructions by utilizing the WSDB-based communication system such as the TVWS system. The WSDB-assisted coverage hole detection system presented below can help reduce the cost on cell planning by limiting the routine drive test to certain areas only, as indicated in Figure 4.34, instead of to the global one.

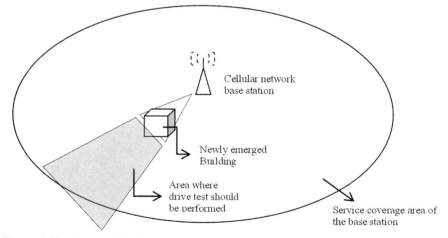

Figure 4.34   Concept of limiting the area for drive tests.

At first glance, it seems that the above problem can also be addressed by wireless positioning approaches. However, either the state-of-the-art wireless LAN (WLAN) positioning or the cellular positioning technology can only locate the active object, which is able to either receive and analyze the signals emitted from base stations to determine its own position or emit signals that will be analyzed by base stations to determine its location. The mentioned obstructions inducing coverage holes are not belonging to active objects, and thus cannot be located by these positioning approaches.

The positioning approach for passive objects, such as relying on the reflection of emitted light beams to determine the location of a passive object, often employs precision measurements where sufficient pairs of emitters and receivers are required, and the placement of each emitter and receiver is also important. Obviously this approach is not suitable for the cellular planning application that requires only the rough position of the emerged obstruction in long distance. Also, this approach needs to build another system rather than the existing wireless communication systems, and hence will increase the cost that the operator may not want to pay.

WSDB REM can be used as a feedback system that is able to indicate in which area the coverage hole is emerged. With this information, the operator (of cellular networks) can conduct drive tests or other measurement-related actions within that area in a very efficient and cost-saving way. The area indicated by the WSDB-assisted coverage hole detection system, such as the gray trapezoid area in Figure 4.35, is named as the "focal area."

The constitution of the WSDB-assisted coverage hole detection system employing REM is depicted as in Figure 4.35. The system shall at least contain two WSDs, a WSDB, an administration centre, and a cellular network base station. The deployed WSDs shall have the geolocation capability (be aware of its

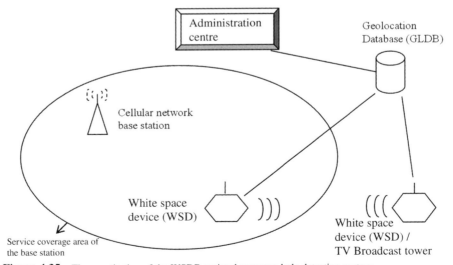

Figure 4.35   The constitution of the WSDB-assisted coverage hole detection system.

geolocation) and be connected to a WSDB. Among the WSDs, there shall be at least one WSD that is able to measure the signal quality of the signal emitted from another WSD. The WSDs shall periodically send back its geolocation and obtained signal quality results to the WSDB. The administration centre shall be able to retrieve the information from the WSDB, and plan drive tests or other measurement-related actions for the base station, based on the retrieved information. In another scenario, one of the two WSDs can be a TV broadcast tower, and the other WSD will report back its location and the receiving signal strength of the TV broadcast tower to the WSDB. In the scenario of TV broadcast tower, there are usually many WSDs that are receiving signal from the TV broadcast tower and reporting back to the WSDB.

There are four steps being used to determine the focal area, which are given below:

(1) *Identify and Record the Connections Between WSDs.*

If a WSD can measure the signal quality of another WSD, then there is a connection between these two WSDs. The connection does not have to be bidirectional, or have real traffic on it. In this step, the WSDB gathers the geolocation information and signal quality report sent by each WSD. After that the WSDB identifies the connections among all WSDs, and records the signal quality for each connection.

(2) *Determine the Deteriorated Connections.*

When a new obstruction (e.g., building) emerged, the signal quality on several connections may drop. If the dropping is significant and persistent, the connection will be considered as a deteriorated connection. To facilitate the judgment, three configurable parameters are used: Δh, nwindow, and nlast. Δh is the dropping threshold of the signal quality, nwindow decides the size of an observation window, and nlast defines the reports received from the last three times. For example, under the configuration where $\Delta h = 10$ dBm, nwindow $= 8$, and nlast $= 3$, any connection whose averaged RSSI calculated from its last three reports is 10 dBm below its averaged RSSI calculated from the previous five reports will be considered as a deteriorated connection.

(3) *Generate the Skeleton Points on the Deteriorated Connections.*

In this step, the skeleton points are generated at which the deteriorated connections cross each other. For those deteriorated connections having no skeleton point on them, just generate the skeleton points on the midpoints of them.

(4) Form the Focal Area Based on the Locations of the Base Station and Skeleton Points

Based on the locations of skeleton points, the focal sector can be determined (with angle θs) by connecting the base station with the two skeleton points which make the angle θs largest. After that the focal sector is further expanded by increasing the angle by $2 \times -\Delta\theta$ (increase $\Delta\theta$ on each side).

Finally the focal area for drive tests is obtained by removing the area within the radius L_{max} from the focal sector, where L_{max} is the distance from the base station to the skeleton point (among the two skeleton points) that has longer distance than the other does.

The above is an example of using WSDB REM for coverage hole detection. The detailed algorithm may vary depending on the need. The WSDB-assisted coverage hole detection system is capable of detecting the coverage holes that exist in the cellular networks, and therefore can significantly save manpower cost for patrolling around to detect coverage holes. The WSDB-assisted coverage hole detection system leverages the existing TVWS networks without introducing new infrastructure for coverage hole detection, and therefore is a very cost-effective solution.

4.5   WHITE SPACE DEVICES

A WSD is a radio working in TVWS band with the capability of opportunistically accessing vacant TV channels. Compared with the radios working in other frequency bands, there are two major differences. One is the more stringent out-of-band (OOB) emission requirement and another is the interface to WSDB to obtain the channels prior to transmission.

A WSD has to meet strict OOB limits, which is required by the regulator for protecting PUs, often in the order of 20–30 dB more stringent than other frequency band radios. Thus, a high-power WSD usually includes a powerful OOB filter to suppress OOB. Besides the strict OOB emission limits, a WSD has to interface to WSDB for obtaining TVWS information.

A WSD has to be certified by the telecom regulator according to TVWS regulation and standard. So far, the mostly prominent TVWS regulations/standards are FCC part 15 subpart H and ETSI EN301598.

There are two possible design approaches to the radio architecture:

One is to make use of the existing radio and its protocol, for example, IEEE 802.11g Wi-Fi radio and add the TV band up/down converter plus the OOB filter in front of the existing radio (Figure 4.36). Note that the RF bandwidth of the existing radio maybe different from a TV channel bandwidth. An IEEE 802.11g Wi-Fi

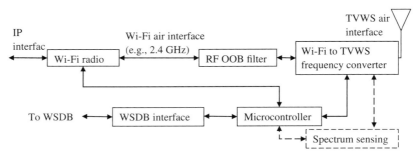

Figure 4.36   Illustration of up/down conversion WSD.

radio has RF bandwidth of 5, 10, or 20 MHz depending on different OFDM symbol rate setting. The lower symbol rate results in a lower data rate. On the other hand, a TV channel has a RF bandwidth of 6, 7, or 8 MHz depending on different countries and whether it is VHF or UHF. Obviously, one TV channel cannot contain a full speed IEEE 802.11g Wi-Fi radio that requires 20 MHz of RF bandwidth. The Wi-Fi radio needs to be set to 5 MHz RF bandwidth in order to be contained in one TV channel. Alternatively, the WSD can also bond two or three adjacent TV channels to allow full speed of 20 MHz signal transmission. This relies on the availability of TVWS channels indicated in the WSDB. This architecture is quick and dirty but can be quickly designed and implemented for TVWS communications. However, its shortcomings are also obvious. First, its spectrum utilization rate is low, because the signal bandwidth is lower than TV channel bandwidth. Second, it is very difficult to carry out channel aggregation among nonadjacent channels with a single transceiver to accommodate the increased data rate. Third, the OOB filter in this architecture employs a RF analogue filter, which is more complex and costly but not as sharp as a digital baseband OOB filter.

Another approach is to design the radio architecture for both baseband and RF (Figure 4.36) In the baseband portion, the digital signal will be packed and framed, modulated according to the available frequencies and bandwidth. It is possible to achieve multichannel aggregation among nonadjacent TV channels by programming OFDM modulation such as setting nulls in certain subcarriers. The digital OOB filter will be integrated or even jointly designed with the OFDM symbols. The TVWS OOB mask can be met easier than the RF OOB filter. The shortcomings of this design are a longer design period and more costly as both baseband and RF designs are required.

Note that the spectrum sensing in Figures 4.36 and 4.37 is for the WSD to select the best TVWS channel that has the lowest co-channel interference from the other WSDs. The basic approach for the WSD to obtain TVWS information is WSDB. Spectrum sensing is the advanced module in the WSD to avoid mutual interference among WSDs. It may not be available in all WSDs.

There are currently a handful of WSDs available in the market. The following is a brief description of the more popular ones.

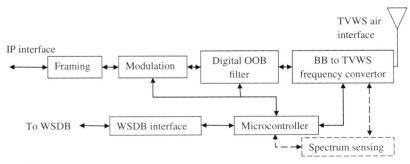

Figure 4.37   Illustration of fully designed WSD.

Figure 4.38 KTS Agility White Space Radio. (*Source*: http://www.ktswireless.com/awr-downloads/. Reproduced with permission of KTS Wireless.)

4.5.1 KTS Wireless AWR

KTS designs and manufactures TV band wireless communication devices for both military and commercial uses. Its commercial TVWS device is called Agility White Space Radio (AWR). The AWR supports the UHF band. The UHF AWR has been approved for working as a fixed TV WSD under FCC TVWS rule Part 15 subpart H. This radio can be used to support video surveillance, SCADA, and broadband wireless Internet access. As for networking, the AWS could be configured as point-to-point (PTP), point-to-mulitipoint (PTM), or simplex (one-way) topologies to carry IP traffic including voice or video. The AWR can connect to FCC-approved WSDB operated by Spectrum Bridge, Inc (SBI). The WSDB drivers are integrated in AWRs so that the operating frequencies of all AWRs can be automatically determined by the devices from the list of available channels obtained from WSDB through the Internet. Figure 4.38 is the picture of AWR. Its specifications are as follows:

Electrical	
RF module frequency (Model AWR-US-U-100)	470 – 698 MHz (TV Channels 14–51)
RF transmit power	10–21.5 dBm with ATPC
Noise figure	4 dB

Spurious & harmonic emissions	FCC Part 15 compliant
Data rate	3.1 Mbps
Modulation	SOQPSK
Channel bandwidth	6 MHz
Frequency selection	100 kHz steps
Selectivity	>60 dB
Operating mode	Half-duplex or simplex
User port	Ethernet (10/100 BT)
Mechanical	
Dimensions	$3.15'' \times 6.6'' \times 1.5''$
Enclosure material	AL, TCP coating
Weight (w/o mount)	1.2 lb
Mounting	Hub Optional bracket for DIN rail or wall Remote/ Subscriber $1'' - 2''$ pole mount bracket
Connectors	
Antenna	Type N (F)
Ethernet/power	(12 to 24 VDC) Watertight RJ45
Environmental	
Operating temperature range	−30 to 60 °C
Operating humidity	As encountered in outdoor mounting
Power	
Input voltage	12 to 24 VDC, via Ethernet cable (PoE)
Current	(12 VDC) 0.4 A (RX), 1.6 A (TX)

4.5.2   6Harmonics Core Adaptive Radio

6Harmonics is another WSD manufacturer. It is a Canada-based communication device company. Figure 4.39 shows 6Harmonics' WSD. This device is certificated by FCC in 2014. 6Harmonics is proposing an adaptive radio networks (ARN) solution for dynamically using the UHF band. The basic specifications of 6Harmonics' WSD are as follows:

Frequency range:	470–770 MHz
RF bandwidth:	5 MHz/6 MHz/8 MHz
Max data rate:	20 Mbps per channel

4.5.3   Carlson's RuralConnect WSD

Carlson Wireless Technologies is a global TVWS technology developer. Its Rural-Connect™ WSD has been certified in 2014 by FCC. It also meets the ETSI TVWS standard. The WSD works in UHF TV channels and provides broadband data to

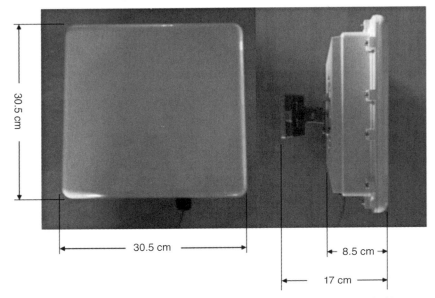

Figure 4.39   6Harmonic Core Adaptive Radio. (*Source*: 6Harmonics Inc. Reproduced with permission of 6Harmonics Inc.)

remote areas. Due to the strong propagation the signal in this frequency band, RuralConnect aims to provide deep coverage for long distance, non-line-of-sight (NLOS), and Internet connectivity to under-served communities over hills and through foliage. Figure 4.40 shows the picture of the RuralConnect WSDs, including outdoor and indoor devices. Its specifications are as follows:

Figure 4.40   Carlson RuralConnect WSDs. (Reproduced with permission of Carlson Wireless.)

Electrical:
Frequency range: UHF 470–698 MHz (US); UHF 470–790 MHz (ETSI)
Channel spacing: 6 MHz (US), 8 MHz (ETSI)
Modulation: 16QAM; QPSK; BPSK
OTA data rates: (6 MHz) 4, 6, 8, 12, and 16 Mbps (8 MHz) 5, 7.5, 10, 15, 20 Mbps
Data rate control: Adaptive or fixed(Receive Interface Proprietary technology is used
                            to reduce co-channel interference)
RX sensitivity: −93 dBm for 10-6 BER using QPSK ½; −86 dBm for 10-6 BER
                            using 16QAM ½; 80 dBm for 10-6 BER using 16QAM
RX blocking   −50 dBm TV transmission on channels $N + 2$; −20 dBm cellular
   resistance: station transmissions
RX max signal: −16 dBm with full linearity
Operating mode: TDD (Time Division Duplexing)
Environment:
Operating temperature: −30 to 55 °C
Operating humidity: Up to 95%, noncondensing
Shock and vibration: MIL-STD-810

4.5.4 WSD From Singapore Power Automation and I²R

Singapore-based Power Automation manufactures the WSDs in collaboration with the Institute for Infocomm Research (I²R). There are Indoor and Outdoor devices. Figure 4.41 shows the first-generation WSDs. I²R has developed different WSDs covering infrastructure mode and mesh networking mode. The developed WSDs have been deployed in Singapore and Philippines in practical applications.

(a)   (b)

Figure 4.41   The first-generation WSDs developed by I²R. (Courtesy of I2R Corporate Communications.)

The WSD's specifications are as follows:

Frequency range	630–750 MHz
Max. P1dB @ 1.5 Mbps	+31 dBm
Channel bandwidth	20/10/5 MHz
Receiver sensitivity	−96 dBm
Data rates	1.5–54 Mbps
RF interface	N-Connector
Data interface	Ethernet (PoE)
Power rating	6 W
Power supply	Passive PoE(24V-48V)

4.5.5 Adaptrum ACRS

Adaptrum ACRS is a WSD certified by FCC for Part 15 Subpart H. It is an Orthogonal Frequency-Division Multiple Access-based (OFDMA) WSD. The manufacturer also announced a feature of reducing potential interference among WSDs, whether the AP antennas are colocated on the same tower or the client devices are in close proximity to one another. ACRS uses spectrum sensing technology to measure radio background noise and the interference level of the entire UHF TV band. These data are used by the operator to select the best channels at individual device locations. The spectrum sensing capability can also assist in the overall network frequency planning.

Other features of ACRS include low-latency, low-frame overhead for higher link throughput.

Adaptrum ACRS is shown in Figure 4.42.

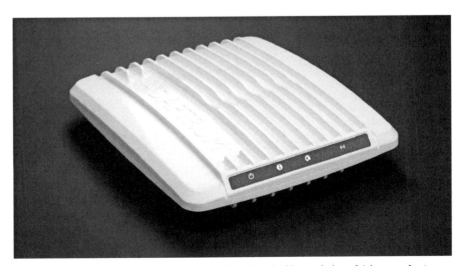

Figure 4.42 Adaptrum ACRS TVWS device. (Reproduced with permission of Adaptrum, Inc.)

The specifications of Adaptrum ACRS are as follows:

Technology	TDD OFDMA
Network topology	Point to Multi-Point
Mobility support	Nomadic & Vehicular up to 40 m/s
Frequency range	400 MHz – 1000 MHz
Channel bandwidth	Flexible up to 10 MHz
	Supporting 6, 7,
	8 MHz TV channels
Aggregated layer 2 data rate	1 – 16 Mbps
(Uplink + Downlink)/6 MHz	
Uplink/downlink ratio	Any
	e.g. $1:1$, $1:3$, $4:1$
Transmit power	100 mW conducted
	(1 W with external amplifier)
Channel bandwidth efficiency	94%
Adjacent channel emission	< -55 dBc
Receiver sensitivity	SNR Data throughput
(6 MHz channel)	
-98 dBm	3.5 dB 2.7 Mbps (QPSK 1/2)
-90 dBm	11.5 dB 7.1 Mbps (16 QAM 2/3)
-81 dBm	20.5 dB 12 Mbps (64 QAM 3/4)
Integrated antenna	PCB Bowtie, Vertically polarized, 0 dBi
Optional external antenna connector	SMA female
Data/Control	10/100 Ethernet
Power supply	48 V PoE
Physical dimensions	$8.5 \times 7.5 \times 1.5$ in^3
Weight	3.6 lb

4.5.6 Redline RTG Connect-IWS

Redline's RTG Connect is for reliable wireless transport for sites featuring TCP/IP-ready and serial SCADA devices and smart metering and automation equipment. There are different versions for different frequency bands. RTG Connect-IWS is the version working in UHF TVWS. Connect-IWS shown in Figure 4.43 is compliant with FCC Part 15 Subpart H.

Figure 4.43   Redline RTG Connect-IWS. (Reproduced with permission of Redline Communications.)

Its specifications are as below:

Frequency range	470–698 MHz
Max. Tx power	+30 dBm
Channel bandwidth (MHz)	0.875/1.25/1.75/2.5/3.5/5/6/7/10/12/14/20
Modulation	BPSK/1/2 to 64QAM5/6, 256QAM
Data rates (Mbps)	3–186.6 Mbps
Security	AES 128/256 (OTA, FIPS 197 compliant);
	HTTPS (SSL), SSH (CLI), SNMP v3; MAC-based, mutual authentication;
	ECDSA Certificates Authentication1; FIPS-140-21
Connection	2xRF TNC(f), USB, 5xRJ-45, DC power
Power	<17 W
Temperature	−40 to 70 °C
Humidity	100% humidity, condensing
Enclosure	IP40

4.5.7 MELD F-Class

MELD has developed another kind of WSD based on broadcasting technology but works in TVWS. The F-class device, MT300AVF, is compliant with FCC TVWS regulation including the emission limits and WSDB interface. The F-class transmitter and matched receiver are shown in Figure 4.44. The receiver can be a set-top

Figure 4.44   MELD's F-class TVWS devices. (*Source*: http://www.meldtech.com/staging2/. Reproduced with permission of MELD Technology, Inc.)

box connected to a conventional TV set, or a dongle device, MTRXm, attached to a smart phone. We note that the F-class WSD is now in Beta version. The emission power of MT300AVF is up to 4W EIRP. MELD's MT600 also implements sub-channeling that supports the same subchannel function as in DTV.

The basic specifications of MT300AVF are as follows.

TX power	Up to +36 dBm EIRP
Resolution	SD/HD
Encoding format	MPEG2
Smart phones supported	IOS (4S, iPads), Android (Galaxy 4S, . . .)

4.5.8 Others

Besides the devices mentioned above, there are other devices designed by manufacturers such as Neul (acquired by Huawei) from the United Kingdom. However, only limited information is available on those devices.

4.6 SUMMARY

In this chapter, we discussed the TVWS technology starting from the general requirements that are specifically for TV band devices. Our discussion spans from the physical layer all the way to the application layer.

In the physical layer, we discussed spectrum sensing and identification issues and solutions. Due to the large wavelength span of TVWS signals, the size of an efficient antenna for a compact WSD is a problem. To address this problem, we presented a unique solution making use of the circuit module itself as the antenna so that the size of the radio is greatly reduced. Furthermore, we discussed the channel aggregation and solutions, OOB leakage control with solutions, and how to utilized TV signal for self-positioning without GPS.

In the MAC layer, we discussed the spectrum sharing solutions for TVWS. we analyzed co-channel interference and coexistence not only between PUs and SUs but also among SUs. For variable bandwidth SU networks, we also presented a dynamic spectrum assignment policy based on optimization theory.

In the network layer, we focused on the reliability of communications. Unlike the wireless communication network over a licensed frequency band, TVWS channels' availability is time-varying. A large area TVWS wireless network must have the capabilities of self-healing and reducing the effect when the availability of its working channel changes due to PUs' operations. To address this issue, we presented a dynamic path switching network routing algorithm to reduce the effect on PUs' operation. In the application layer, we presented a few WSDBs with advanced functions. These functions include area-based query for spectrum usage, broadcast WSDB without Internet access for higher security, and WSDBs with high-priority channels and higher QoS users. We also presented some novel applications of

TVWS systems. These include how the information collected by WSD networks could be used to build a radio environment map for other wireless systems, and how the information of WSD networks can be used to help telecommunication providers to detect coverage holes.

For more information on future TVWS developments, readers can refer to Chapter 7.

REFERENCES

1. G. M. Galvan-Tejada, M. A. Peyrot-Solis, and H. J. Aguilar, *Ultra Wideband Antennas: Design, Methodologies, and Performance*. CRC Press, June 24, 2015.
2. R. C. Johnson, *Antenna Engineering Handbook*, 3rd edn, McGraw-Hill, Inc.
3. C. Leray, Mobile telephone with an integrated component in the antenna volume. European Patent Application EP1317116, 2003.
4. B. Lindell, Placing of components on an antenna arrangement. US Patent US7250911, 2007.
5. Y. Ban, J. Chen, S. Yang, J. Li, and Y. Wu, Low-profile printed octa-band LTE/WWAN mobile phone. *IEEE Trans. Antennas Propag.*, 61 (7), 3889–3894 (2013).
6. S. Chien, H. Chen, C. Su, F. Hsiao, and K. Wong, Planar inverted-F antenna with a hollow shorting cylinder for internal mobile phone antenna. *IEEE Antennas Propag. Soc. Int. Symp.*, 2, 1947–1950 (2004).
7. P. Vainikainen, J. Holopainen, and M. Kyro, Antennas for digital television receivers in mobile terminals. *Proc. IEEE*, 100 (7), 2341–2348 (2012).
8. M. Subhedar and G. Birajdar, Spectrum sensing techniques in cognitive radio networks: a survey. *Int. J. Next Generation Networks*, 3 (2) (2011).
9. C. Ghosh, S. Roy, and D. Cavalcanti, *Coexistence challenges for heterogeneous cognitive wireless networks in TV white spaces. IEEE Wirel. Commun.*, 18 (4), 22–31, (2011).
10. M. Nekovee, Cognitive radio access to TV white spaces: spectrum opportunities, commercial applications and remaining technology challenges, in IEEE symposium on New Frontiers In Dynamic Spectrum, IEEE, April 2010.
11. D. X. Brown, An analysis of unlicensed device operation in licensed broadcast service bands, in Proc. IEEE Int. Symp. Dynamic Spectrum Access Netw. (DySPAN'05), Nov. 2005, pp. 11–29.
12. FCC , Second Report and Order and Memorandum Opinion and Order, Report ET Docket No. 08–260, November 2008.
13. R. Rajbanshi, A. M. Wyglinski, and G. J. Minden, An efficient implementation of NC-OFDM transceivers for cognitive radcios, in 1st International Conference on Cognitive Radio Oriented Wireless Networks and Communications, 2006.
14. D. Noguet, M. Gautier, and V. Berg, Advances in opportunistic radio technologies for TVWS. *EURASIP J. Wirel. Commun. Netw.*, 2011.
15. V. Berg, Zs. Kollar, R. Datta, P. Horvath, D. Noguet, and G. Fettweis, Low ACLR communication systems for TVWS operation, in Conference Proceedings of Future Network & MobileSummit, 2012.
16. G. Fettweis, M. Krondorf, and S. Bittner, Gfdm: generalized frequency division multiplexing, in IEEE 69th Vehicular Technology Conference, 2009., VTC Spring 2009. April 2009, pp. 1–4.
17. A. Makki, A. Siddig, M. Saad, J. R. Cavallaro, and C. J. Bleakley, Indoor localization using 802.11 time differences of arrival. *IEEE Trans. Instrum. Meas.*, 2015.
18. M. Rabinowitz and J. Spilker, A new positioning system using television synchronization signals. *IEEE Trans. Broadcast.*, 51 (1), 51–61 (2005).
19. X. Wang, Y. Wu, and J.-Y. Chouinard, A new position location system using DTV transmitter identification watermark signals. *EURASIP J. Appl. Signal Process.*, 2006 (42737), 1–11 2006.
20. L. Dai, Z. Wang, C. Pan, and S. Chen, Wireless positioning using TDS-OFDM signals in single-frequency networks. *IEEE Trans. Broadcast.*, 58 (2), 236–246 (2012).

21. D. Serant, O. Julien, C. Macabiau, P. Thevenon, S. Corazza, M. Dervin, M.-L. Boucheret, and L. Ries, Development and validation of an OFDM/DVB-T sensor for positioning, in Proc. IEEE/ION Position Location and Navigation Symp., May 2010, pp. 988–1001.

22. J. Yang, X. Wang, M. J. Rahman, S. I. Park, H. M. Kim, and Y. Wu, A new positioning system using DVB-T2 transmitter signature waveforms in single frequency networks. *IEEE Trans. Broadcast.*, 58 (3), 347–359 (2012).

23. R. Sawai, R. Kimura, N. Sato, and G. Xin, Coexistence mechanism and its algorithm, Document IEEE 802.19-10/0145r1, October 2010.

24. FCC , Third memorandum opinion and order, Report ET Docket No.12-36, April 2012.

25. W. Daniel and H. Wong, Propagation in Suburban Areas at Distances Less than Ten Miles, FCC, Washington, DC, Report OET TM 91-1, January 1991.

26. Y. Song, C. Zhang, and Y. Fang, Multiple multidimensional knapsack problem and its applications in cognitive radio networks, in Proc. IEEE Military Commun. Conf (MILCOM'08), November 2008, pp. 1–7.

27. S. Martello and P. Toth, *Knapsack Problems: Algorithms and Computer Implementations*, John Wiley & Sons, Ltd., 1990.

28. A. Bourdena, G. Kormentzas, G. Mastorakis, E. Pallis, P. Marques, J. Mwangoka, J. Rodriguez, J. Kubasik, H. Bogucka, M. Parzy, G. Schuberth, L. Gomes, and H. Alves, Dynamic radio resource management algorithms for an efficient use of TVWS, COGEU FP7 ICT-2009.1, Tech. Rep., January 2011.

29. A. Bourdena, E. Pallis, G. Kormentzas, C. Skianis, and G. Mastorakis, QoS provisioning and policy management in a broker-based CR network architecture, in Proc. IEEE Global Commun. Conf. (GLOBECOM'12), December 2012, pp. 1841–1846.

30. A. Bourdena, G. Kormentzas, E. Pallis, and G. Mastorakis, A centralised broker-based CR network architecture for TVWS exploitation under the RTSSM policy, in Proc. IEEE Int. Conf. Communications(ICC'12), 2012, pp. 5685–5689.

31. D. Chen, Q. Zhang, and W. Jia, Aggregation aware spectrum assignment in cognitive ad-hoc networks, in Proc. Int. Conf. Cognitive Radio Oriented Wireless Netw. and Commun. (CrownCom'08), May 2008, pp. 1–6.

32. C. Li, W. Liu, J. Li, Q. Liu, and C. Li, Aggregation based spectrum allocation in cognitive radio networks, in IEEE/CIC Int.l Conf. Commun. in China, August 2013, pp. 50–54.

33. X. M. H. Ma, L. Zheng, and Y. Luo, Spectrum aware routing for multi-hop cognitive radio networks with a single transceiver, in 3rd International Conference on Cognitive Radio Oriented Wireless Networks and Communications, CrownCom, 2008.

34. W. L. Geng Cheng and W. C. Yunzhao Li, Spectrum aware on-demand routing in cognitive radio networks, in IEEE International Symposium on New Frontiers in Dynamic Spectrum Access Networks, 2007.

35. M. F. K. Chowdhury, SEARCH: a routing protocol for mobile cognitive radio Ad-Hoc networks. *Comput. Commun.*, 2009.

36. A. G. M. Zhu and I. G. S. Kuo, STOD-RP: a spectrum-tree based on-demand routing protocol for multi-hop cognitive radio networks, IEEE GLOBECOM, 2008.

37. M.-H. Tao, S. W. Oh, Yugang Ma, Dynamic path switching routing protocol for cognitive radio networks, in IEEE Wireless Communications and Networking Conference (WCNC 2015), New Orleans, 2015.

38. K. Habak, M. Abdelatif, H. Hagrass, K. Rizc, and M. Youssef, A Location-aided routing protocol for cognitive radio networks, in International Conference on Computing, Networking and Communications (ICNC), 2013.

39. W. Kim, M. Gerla, S. Y. Oh, K. Lee, and A. Kassler, CoRoute: a new cognitive anypath vehicular routing protocol. *Wirel. Commun. Mobile Comput.*, 11 (12), 1588–1602 (2011).

40. X. Tang, Y. Chang, and K. Zhou, Geographical oportunistic routing in dynamic multi-hop cognitive radio networks, in *IEEE 2012 Computing, Communications and Applications Conference*, 2012, pp. 256–261.

41. I. Filippini, E. Ekici, and M. Cesana, Minimum maintenance cost routing in cognitive radio networks, in MASS, 2009, pp. 284–293.

42. X. Huang, D. Lu, P. Li, and Y. Fang, Coolest path: spectrum mobility aware routing metrics in cognitive Ad Hoc networks, in 31st International Conference on Distributed Computing Systems (ICDCS), June 2011, pp. 182–191.

43. IETF , Protocol to Access White-Space (PAWS) Databases. May 2015.

44. FCC , Notice of Proposed Rule Making and Order, Report ET Docket No. 03–322, December 2003.

45. FCC , Second Memorandum Opinion and Order, Report ET Docket No. 10–174, September 2010.

46. A. Ghasemi, and E. S. Sousa, Spectrum sensing in cognitive radio networks: requirements, challenges and design trade-offs cognitive radio communication and networks. *IEEE Commun. Mag.*, 46 (4), 32–39 (2008).

47. S. Mangold, A. Jarosch, and C. Monney, Operator assisted cognitive radio and dynamic spectrum assignment with dual beacons: detailed evaluation, in First International Conference on Communication System Software and Middleware, 2006.

48. FCC , First report and order and further notice of proposed rulemaking, October 2006.

49. S. W. Oh, The path of TV white space, in IET International Conference on Wireless Communications and Applications. 2012.

50. S. W. Oh and C. C. Chai, Geo-location database with support of quality of service for TV white space, in IEEE International Symposium on Personal, Indoor and Mobile Radio Communications. 2013.

51. X. Feng, Q. Zhang, and J. Zhang, A hybrid pricing framework for TV white space database. *IEEE Trans. Wirel. Commun.*, 13 (5), 2626–2635 (2014).

52. V. Erceg et al., An empirically based path loss model for wireless channels in suburban environments. *IEEE J. Sel. Area. Commun.*, 17 (7), 1205–1211 (1999).

53. K. T. Herring, J. W. Holloway, D. H. Staelin, and D. W. Bliss, Path-Loss Characteristics of Urban Wireless Channels. *IEEE Trans. Antennas Propag.*, 58 (1), 171–177 (2010).

54. T. S. Rappaport, *Wireless communications: principles and practice*, vol. 2, Prentice Hall PTR, New Jersey, 1996.

55. M. Riback, J. Medbo, J.-E. Berg, F. Harrysson, and H. Asplund, Carrier frequency effects on path loss, in Proc. IEEE 63rd Veh. Technol. Conf. (VTC), May. 2006, pp. 2717–2721.

56. M. B. Kjærgaard, A taxonomy for radio location fingerprinting, in Proc. Int. Symp. Location and Context-Awareness (LoCA), 2007, pp. 139–156.

57. P. Bahl and V. N. Padmanabhan, Radar: an in-building RF-based user location and tracking system, in Proc. IEEE Conf. Computer Commun. (INFOCOM), Mar. 2000, pp. 775–784.

58. A. LaMarca, Y. Chawathe, S. Consolvo, J. Hightower, I. E. Smith, J. Scott, T. Sohn, J. Howard, J. Hughes, F. Potter, J. Tabert, P. Powledge, G. Borriello, and B. N. Schilit, Place lab: device positioning using radio beacons in the wild, in Proc. Int. Conf. Pervasive Computing, May 2005, pp. 116–133.

Chapter 5

Worldwide Deployment

Conventional communication systems and regulatory policies encourage exclusive licensing of spectrum to specific operators. Static spectrum allocation is driven by the need to minimize interference and serve as dedicated broadband access for mobile services that already created a huge telecommunication ecosystem. However, even though 2.5 billions of the global population are broadband connected, another 5 billion people are not and it does not seem that those people will be able to get connected soon [1]. It is believed that spectrum sharing with dynamic/opportunistic spectrum access will be able to address the need for affordable access. Dynamic access to the spectrum is on its way, bringing hope that more people will gain access to more affordable broadband services.

In order to investigate how much additional wireless spectrum is required to support the increasing demand on mobile services and applications, spectrum occupancy measurement campaigns have been launched worldwide, as shown in Figure 5.1. It is found from the results of these campaigns that the availability of spectrum is not constant in many allocations. Therefore, the potential for spectrum sharing is quite big and pivotal [2].

Spectrum sharing requires the deployment of communication systems that, unlike conventional communication systems, use the radio resources opportunistically without interfering with the primary users while still maintaining the equality of access among secondary users. The resource scheduling is not managed by specific telecom operators or spectral owners; instead, other techniques are applied to access the sharing spectrum. TVWS-based spectrum sharing technology suggests maintaining a database, where the spectrum reservations can be done by sending requests to the database [3].

To successfully develop, demonstrate, and deploy the TVWS and dynamic spectrum access technology, several TVWS trials/projects have been completed or are ongoing,[1] and corresponding test beds have also been established. Figure 5.2 shows that the TVWS technology is already being widely deployed in projects occurring on five continents. With the involvement of governments, regulators, industry, and academia, there are many activities and efforts in the direction to

---

[1] The Dynamic Spectrum Alliance—Worldwide Trials and Pilots Web Site, http://www .dynamicspectrumalliance.org/pilots.html.

TV White Space: The First Step Towards Better Utilization of Frequency Spectrum, First Edition.
Ser Wah Oh, Yugang Ma, Ming-Hung Tao, and Edward Peh.
© 2016 by The Institute of Electrical and Electronics Engineers, Inc. Published 2016 by John Wiley & Sons, Inc.

Figure 5.1 The spectrum occupancy measurement campaigns [2].

move the wireless communication to a new stage where the efficiency and flexibility on spectrum usage can be extremely high.

These deployments have taken place in locations as diverse as the United States, the United Kingdom, South Africa, Singapore, Japan, South Korea, Ghana, the Philippines, Kenya, Tanzania, Uruguay, and Malawi. Each of these projects has been launched with the support and authorizations from the relevant regulatory authorities. In all of these projects, additional bandwidth for Internet access can be achieved without causing interference to incumbent users, including broadcasters. Some of the deployments have provided Internet access to primary and secondary schools, university campuses, libraries, community centers, health care facilities, government offices, and small and medium sized businesses.

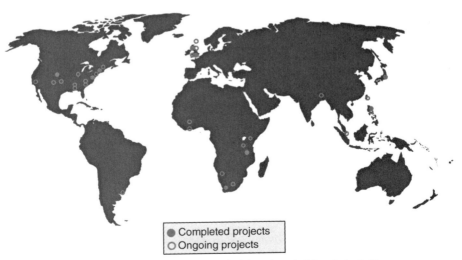

Figure 5.2 The worldwide deployment of TVWS trials and test bed (see footnote 1).

These projects demonstrate that the feasibility of harvesting new sub-gigahertz spectrum in TV white space bands. These new bands could provide an affordable wireless solution to extending the reach of broadband wireless network to rural or underdeveloped areas, and could also enable the Internet access under the long-range or even the indoor communication scenario.

5.1 NORTH AMERICA

Table 5.1 summarizes the TV white space trials in North America with the details described in the following sections.

Table 5.1 TV White Space Trials in North America

Service Provider	Devices	Trial Location	Trial Application
Axiom	Adaptrum	Washington County, Maine	Wireless broadband services
CVALINK Broadband	Adaptrum	Louisa County	Wireless broadband services
Gigabit Libraries Network, University of Delaware	Carlson Wireless, KTS Wireless, Adaptrum	Sussex County, Delaware	Digital service centre
Gigabit Libraries Network, University of New Hampshire's Broadband Center of Excellence	Carlson Wireless, KTS Wireless, Adaptrum	Durham, NH	Research and development
Gigabit Libraries Network, Humboldt County library	Carlson Wireless, KTS Wireless, Adaptrum	Humboldt County, CA	External Wi-Fi backhaul
Gigabit Libraries Network, Delta County library	Carlson Wireless, KTS Wireless, Adaptrum	Delta County, CO	External Wi-Fi backhaul
Gigabit Libraries Network, Skokie Public Library	Carlson Wireless, KTS Wireless, Adaptrum	Skokie, IL	External Wi-Fi backhaul
Gigabit Libraries Network, State Library of Kansas	Carlson Wireless, KTS Wireless, Adaptrum	Topeka, KS	External Wi-Fi backhaul
Gigabit Libraries Network, Pascagoula School District	Carlson Wireless, KTS Wireless, Adaptrum	Pascagoula, MS	External Wi-Fi backhaul
Port of Pittsburgh Commission	Adaptrum	Pittsburgh, PA	Wireless broadband services
West Virginia University	Adaptrum	Morgantown, WV	External Wi-Fi backhaul
Spectrum Bridge	6Harmonics	Wilmington, NC	Wireless broadband services
The Gatineau Valley Societe d'aide au développement	6Harmonics	Pointe a David	External Wi-Fi backhaul

5.1.1 The United States

• Washington County

Adaptrum has been working with the local wireless Internet service provider Axiom and initiating a project to provide a TV white space broadband service to the rural areas in Washington County, Maine. Adaptrum's featured ACRS 2.0 technology, certified by the FCC, is employed in the project in June 2014. In the current phase, there are total five base sites and 40 customers across the county.

ACRS 2.0's NLOS OFDMA technology utilized the superior propagation characteristics of the UHF TV bands, and therefore can provide reliable connectivity in the harsh environment within the rural, suburban, and urban areas. The project can provide a link speed of around 10 Mbps for a NLOS distance from the base site to the customer premise, which is 2–5 miles. The next phase of the project includes 600 customer entities located in densely wooded and hard-to-reach areas.

• Louisa County

In Louisa County, CVALINK Broadband, the Internet service provider, partnered with Adaptrum to launch Adaptrum's ACRS 2.0 technology in rural areas, which aims to provide fast and reliable Internet services to the county.

Adaptrum installed a base station on Louisa's water tower and tested the link back to the company's offices on Davis Highway. The equipment demonstrated that TV white space technology is superior in both reachable range and reliability while compared with the side-by-side 900 MHz and 2.4 GHz Wi-Fi-based products.

The deployment eventually enables CVALINK Broadband to provide services to customers in dense wooded areas that are originally unreachable using traditional unlicensed bands and technologies. With the more powerful penetrating ability, TV white space can easily penetrate through trees and therefore can connect more citizens than before. Since CVALINK Broadband partnered with Adaptrum, the county of Louisa has also committed to test the TV white space technology at the office building as it seems to be able to provide broadband Internet services to all its residents (Figures 5.3 and 5.4).

• Gigabit Libraries Network "Libraries Super Wi-Fi" project

The Gigabit Libraries Network (GLN), an open collaboration of innovative libraries testing new technologies, announced the results of its open call to participate in a library consortium to run usability trials across the United States on newly available unlicensed wireless communications spectrum called super Wi-Fi or TV white space. There are several participating library systems and consortia: Delaware, New Hampshire, Humboldt County, CA, Delta County, CO, Skokie, IL, Kansas City, and Pascagoula, MS.

Delaware has been selected as one of the new state pilots in the U.S. State Librarian. The University of Delaware is leading an ambitious state-wide initiative to determine the feasibility of deploying TV white space hubs at every library

Figure 5.3 TV white space base site on Louisa Water tower. (*Source:* http://www.adaptrum.com/acrs2launch/Launch_DeployPics_Louisa.htm#carousel-dev-images. Reproduced with permission of Adaptrum.)

facility in the state. The first two libraries chosen for the initial pilots are the Greenwood Public Library and Laurel Public Library in Sussex County, Delaware. It is believed that white space pilot will extend the existing Delaware Libraries services to further support business growth and entrepreneurship in Delaware.

There is also a state-wide pilot consortium led by the University of New Hampshire's Broadband Center of Excellence (BCOE), which aims to deploy TVWS system in rural parts of the state and hosts a collaboration with GLN called "Libraries Whitespace Lab" to explore the issues as well as the applications relating to the concept "libraries anywhere."

Humboldt County library has also been selected as a test site for the TV white space "Super Wi-Fi" project. This project will test the efficacy of using vacant TV broadcast spectrum as a means of providing reliable wireless

Figure 5.4 Residential client site in Louisa County. (*Source:* http://www.adaptrum.com/acrs2launch/Launch_DeployPics_Louisa.htm#carousel-dev-images. Reproduced with permission of Adaptrum.)

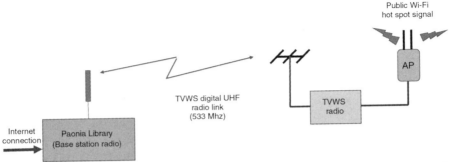

Figure 5.5   System architecture for the library-sponsored public hotspots in Paonia, CO. (Courtesy of John Gavan.)

Internet access for public use. The TV white space will be deployed using Access Humboldt's tower at the Muni-hub, and will deliver wireless connectivity through barriers such as trees, hills, and buildings.

In Delta County, the GLN WhiteSpaces Pilot went to live in Paonia in October 2013. The Delta County library can now provide Wi-Fi signal to the main street in the town with the help of the TV white space relay to fulfill the need to transmit the signal to a park located 1 mile away from the library so that summer festival vendors can use the Wi-Fi for credit card transactions (Figure 5.5 and 5.6). In Delta County, the libraries are the centers of the community. In many cases, the libraries are the only place where people can go to get on the Internet to apply for a job, file a tax return, seek information, do e-mail, browse social media, and do the myriad of things that they can only do on the Internet. With the help of TV white space, people can now access the Internet more easily even in the street. The Whitespaces Pilot in Paonia

Figure 5.6   The Wi-Fi access point installation in the main street of Paonia, CO. (Courtesy of John Gavan.)

concluded [4] that the TV white space technology is especially suitable (1) where other wireless solutions have failed, (2) where heavy tree foliage exists, (3) in hilly areas, (4) in rural, unserved, and underserved areas, (5) where towers are prohibited or impractical, or (6) in areas of low population density.

Skokie Public Library has also been selected as a test site for the Library Super Wi-Fi project. Again the project will test the efficacy of using vacant TV broadcast spectrum as a means of providing reliable wireless Internet access to public spaces in the Village of Skokie.

The State Library of Kansas announces the establishment of the Kansas K20-Librarians Whitespace Pilot. The pilot was initiated by the Kansas City K-20 Librarians Initiative and coordinated by the State Library, under the national GLN's Libraries Super Wi-Fi project. The pilot is working to spread Internet access across the community. This was inspired by the Google Fiber that is being deployed in Kansas City currently. However, Google Fiber cannot reach many areas of the city, so TV white space is an ideal solution to fill in the gaps.

After Hurricane Katrina, the Pascagoula School District wanted to have TV white space technology available as a disaster recovery resource. Therefore, a TV white space project was initiated at Pascagoula School District in March 2014. The community built a mobile unit consisting of a tower and telescope. It can be moved for community events, such as a county fair or a concert, or in the case of emergencies, become a transportable Wi-Fi hotspot. The Pascagoula School District also plans to use the TV white space to replace a DSL connection in an Adult Learning Center, which will triple the bandwidth and lower connection cost.

• Pittsburgh

In Pittsburgh, the Port of Pittsburgh Commission engaged with Adaptrum to conduct the Interoperability Test Bed(ITB) Program that aims to demonstrate and evaluate new technologies and services that can enhance inland waterway safety, transportation efficiency, and environmental systems management. The ongoing trial involves the U.S. inland waterway system, where broadband wireless services contribute to improving the lock and vessel safety, crew liability, and overall regional economic competitiveness.

Two Adaptrum TV white space stations were deployed on the rooftop of the Carnegie Science Center to connect boats on the Monongahela, Allegheny, and Ohio rivers in Pittsburgh (Figure 5.7). Using Adaptrum's ACRS2.0 technology, the project demonstrated how TV white space can effectively connect vessels traveling at different speeds to allow VoIP services and video calls (Figure 5.8). The handover function was also successfully demonstrated with the scenario where the client boats went from one TV white space cell to another TV white space cell. The ability for NLOS transmission enables the signal to span several miles on all three rivers, which shows that the technology is a critical element in creating wireless waterways.

• Morgantown, WV

West Virginia University is the first University in the United States to use vacant broadcast TV channels to provide wireless broadband Internet services

Figure 5.7   Base site at Carnegie Science Center. (*Source:* http://www.adaptrum.com/acrs2launch/Launch_DeployPics_Pittsburgh.htm. Reproduced with permission of Adaptrum.)

for its staffs and students. Using FCC-certified TV white space equipment from Adaptrum, a San Jose start-up developed a software-defined radio that can dynamically switch channels, adjust channel bandwidths, and tune power levels. The TV white space deployment went live in February 2014 and is the first deployment in a U.S. University.

Figure 5.8   ACRS 2.0 client radio on boats. (*Source:* http://www.adaptrum.com/acrs2launch/Launch_DeployPics_Pittsburgh.htm. Reproduced with permission of Adaptrum.)

The key component of this TV white space deployment is the Adaptrum base station, which has been placed on the rooftop of the Engineering Sciences Building near the center of Morgantown. The base station is equipped with the directional antennas, cabling, power supply, performance and availability monitoring equipment, switching facility, and interconnections for backhaul through the University's backbone to the Internet. There are five Adaptrum client devices, one for each personal rapid transit(PRT) station, along with their own antennas, and an Ethernet connection to one or more Wi-Fi access points. Even in the harsh environment for radio transmission, the Internet service is still available for end users. They just need to simply use Wi-Fi to create a connection for laptops, smart phones, and tablets. The Adaptrum client device will then make the necessary conversions at its end and send the packets over the vacant TV spectrum to the base station, which will send the packets to the Internet via a connected route.

Phase 1 provided Wi-Fi backhaul to the on-campus TRAM system with five platforms over a 4 mile loop around the campus, which benefits more than 15,000 riders who travel the campus every day using electric-powered vehicles. It is planned in Phase 2 that additional sites on campus including an off-campus building, nursery, and parking lots will also be connected. Both Phase 1 and Phase 2 have been completed with average NLOS link distances between 1 and 3 miles, which provides link throughput around 10 Mbps.

• Wilmington, NC

In 2012, nearly 3 years after the FCC proposed the first white space commercial network in Wilmington, NC, the city finally implemented this project. The TV white space network that runs through Spectrum Bridge, one of the first FCC-approved white space databases, can deliver remote Internet services to previously hard-to-access areas. Wilmington was chosen because it was the first city that switched from analog to digital TV, and therefore freed up the wireless spectrum for white space. The TV white space network provides broadband access to the New Hanover County Juvenile Center, police stations, and also the Youth Enrichment Zone. Also, the government has been using the TV white network to connect two local parks and several public gardens, and to monitor water levels, water quality, and public lighting. Public security will be expanded with the help of the TV white space network to carry video surveillance and manage traffic congestion.

The TV white space network is part of Wilmington's wider, ongoing "Smart City" initiative, which aims to improve the standard of living and drive the economy by creating a better communications infrastructure. The TV white space network has proved in the deployment to be particularly valuable to TVWS equipment manufacturers, and serves as a test bed for new applications.

5.1.2 Canada

Canada's first TV white space network was built by 6Harmonics in October 2012, alongside the Gatineau Valley Societe d'aide au développement des collective.

Pointe a David was chosen as the trial site because the Internet accessing requirement in that area is increasing day by day. There are over 100 campsites, in which Internet access is difficult due to dense vegetation. Because TV white space technology utilizes lower frequency bands to transmit the signal, the signal is able to penetrate trees, buildings, and rugged terrain more easily.

The Gatineau Valley Societe d'aide au développement invested a total amount of $20,000 into the project, with the hope that the network will be improved significantly.

5.2 EUROPE

5.2.1 The United Kingdom

In April 2013, Ofcom held a stakeholder event to discuss next steps with TV white spaces [5]. At the event, initial proposals for running a pilot of the UK TVWS framework were presented. Stakeholders were encouraged to express their interest if they wished to participate in the pilot. A follow-up stakeholder event was then held on July 25, 2013. to update stakeholders with corresponding pilot developments.

There are several main parts constituting the pilot, which are discussed in the following sections:

5.2.1.1 White Space Database Contracting and Qualification

This part started in December 2013 when the first contracts for TV white space databases were signed. The qualification process of the contracts was started in January 2014 and was completed in May 2014. Currently, there are eight WSDBs that have been qualified (as shown in Table 5.2) for the purposes of the pilot, with the final WSDB being qualified in October 2014. The window for new WSDB applications closed at the end of August 2014.

Table 5.2   Database Qualified in the UK TV White Space Pilot

Database Provider	Date Qualified
Spectrum bridge	May 21, 2014
Nominet	June 10, 2014
NICT (National Institute of Information and Communications Technology)	June 10, 2014
Fairspectrum	June 10, 2014
Google	August 21, 2014
Sony	September 8, 2014
iconectiv	October 8, 2014
Microsoft	October 23, 2014

5.2.1.2 White Space Trials

The first trial was licensed in June 2014 and now there are 11 trials that have been licensed. These trials cover a wide range of applications, including public Wi-Fi, webcam backhaul, rural and maritime broadband services, remote sensing, academic research, digital signage, local broadcasting, and CCTV distribution. The window for new trial applications closed at the end of August 2014. Trials are expected to continue until 2015. Some of the trials potentially will continue until the introduction of commercial operation.

Figure 5.9 shows the location of licensed trials in the United Kingdom. Table 5.3 lists the trials and provides information on the trials, devices used, WSDB partners, and use cases for each of the trials.

Ofcom teamed up with alliance members 6Harmonics, Mediatek, and Google to launch a TV white space trial at the ZSL London Zoo. The trial was using the TV white space technology to stream daily live footage of zoo animals to YouTube.

The equipment provided by 6Harmonics and Mediatek is used to transmit images of meerkats, otters, and tortoises from their habitat to YouTube via

Figure 5.9   The location of the licensed TV white space trials in the United Kingdom.

Table 5.3 TV White Space Trials in the United Kingdom [5]

Service Provider	Database	Devices	Trial Location	Trial Application
Centre for White Space Communications, University of Strathclyde	Spectrum Bridge	6Harmonics	Glasgow	External Wi-Fi and webcam backhaul
Click4Internet	NA	Neul/6Harmonics	Isle of Wight	Land ↔ private boat broadband
Cloudnet Solutions	Fairspectrum	Carlson wireless	Orkney Islands	Land ↔ ferry broadband
CYP (UK) Ltd	Spectrum Bridge	MELD technologies	Shepperton	Digital signage
Google/ZSL London Zoo	Google	Mediatek/ 6Harmonics	London Zoo	Live video feeds of animal enclosures
Kings College (as a collaboration with the Joint Research Centre of the EC and Eurecom)	Fairspectrum/ Spectrum Bridge	Sinecom/KTS wireless, Carlson wireless, Eurecom, Runcom, Interdigital, NICT	London and others	Research and development
Love Hz Ltd and Nominet Ltd	Nominet	Adaptrum	Oxford	Community sensor network (flood detection)
National Institute of Information and Communications Technology (NICT)	NICT	NICT	London	Research and development
Nominet	Nominet	Eurecom	Oxford and KCL Strand	Research and development
Peerless AV	Spectrum Bridge	MELD technologies	Watford	Digital signage

spectrum temporarily licensed by Ofcom. Google provided spectrum database in the trial that aims to help ZSL London Zoo test the technology for efforts to protect endangered animals in the wild (Figure 5.10).

NICT and King's College London have conducted trials within the Ofcom TV white spaces pilots, in which WSDs communicating with WSDBs to obtain the permitted channel/power information using the evolving IETF PAWS standard. PAWS is the protocol to access spectrum databases. Other aspects

Figure 5.10 Meerkats and otters in London Zoo are being broadcast on Youtube using TV white space.

investigated in the trials included the performance of the WSDs, and also coexistence issues (Figure 5.11).

The WSDs used in these trials are in fact a LTE system (3GPP Release 8) extended to operate in TV channels and an IEEE 802.11af system, both are developed by NICT. These WSDs satisfy the requirements of the ETSI 301598 specification, including spectrum mask characteristics. The WSDB was also developed by NICT, and has been qualified for operation in the United Kingdom by Ofcom.

The LTE system enabled mobile broadband communications by using TVWS frequencies and achieved a downlink throughput of 45 Mbps in FDD mode (20 MHz each for uplink and downlink, from aggregating three TV channels) and 19 Mbps in TDD mode (20 MHz, aggregating three TV channels). The IEEE 802.11af system established a 3.7 km point-to-point link using one TV channel, with the downlink throughput that is over 2 Mbps.

Also at the Strand Campus of King's College London, the European Union Joint Research Centre (EU-JRC) completed the testing of the Carlson Wireless Rural Connect TVWS broadband system. The testing is designed to qualify TVWS technology for commercial use in the United Kingdom, especially to test the

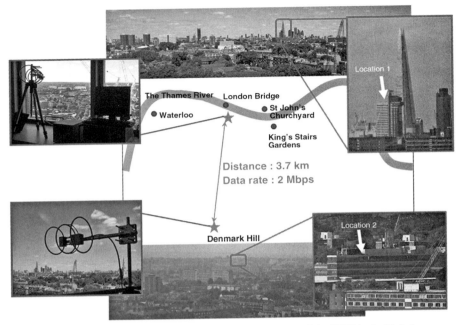

Figure 5.11   Trial locations and system deployments in the trial built by NICT in the United Kingdom. (*Source:* http://www.nict.go.jp/en/press/2014/07/24-1.html. Reproduced with permission of NICT (National Institute of Information and Communications Technology, Japan).)

connectivity using the Carlson gear for in-building, between-building, and long-distance scenarios. The testing was also aiming to verify system communication with a WSDB provided by Fair Spectrum of Helsinki, Finland. The equipment was deployed at campuses of King's College London, located in the center of London. During the testing, the feasibility of TVWS utilization in a metropolitan area was proved.

At the University of Strathclyde, Glasgow, Microsoft together with Mediatek, 6Harmonics, and Aviacomm successfully developed and deployed the world's first triple band Wi-Fi transceivers, which are part of the TV white space project hosted by the Centre for White Space Communications (CWSC). The pilot will use TV white space to fill urban broadband gaps and enable smart city applications. The first phase of the project supported by the Scottish Government and Microsoft will deploy radios at predefined locations around the City Centre Campus inside the University. The CWSC is also considering how to harvest and aggregate data from distributed environmental sensors. These data are used to feed a novel City Observatory and an Open Glasgow portal. A number of countries represented in the Glasgow 2014 Commonwealth Games are also embarking on their own white space journeys.

Despite Scotland has far slower Internet speed than the rest of the United Kingdom, it was beginning to use TV white space to provide Internet connectivity and communications to the Orkney Ferries serving the Orkney Islands and in the

Pentland Firth. The deployment was aiming to provide wireless broadband access for these ships.

It has been a challenge for residents of the Orkney Islands to access the Internet from home or during the daily commute via ferry. Directors Greg Whitton and Nikki Linklater of CloudNet IT Solutions were deploying TVWS gears to provide Internet access at ferry terminals and on the ferries so that passengers are able to access the Internet while traveling.

In April 2011, the UK government's Technology Strategy Board supported a six-partner consortium to work on a rural broadband trial network that used TV white space spectrum to provide broadband communication to a small rural community on the south part of the Isle of Bute, Scotland [6]. The goal was to investigate and demonstrate the potential of the white space spectrum for providing broadband access to remote rural areas in challenging terrain. The 18-month long project involved the planning and installation of white space radio links from the local telephone exchange to eight premises in the surrounding area. The project also established the backhaul connectivity from the telephone exchange to the mainland, which is then connected to the Internet (Figure 5.12).

Cambridge White Spaces Trial launched in June 2011 was designed to evaluate both the technical capabilities of the TV white space technology and the potential requirements of end user applications and scenarios. The trial explored and measured a range of applications, including rural wireless broadband, urban pop-up coverage, and even M2M communication. The trial concluded that TV white spaces can be successfully utilized to help satisfy the rapidly growing demand for wireless accessing. The partners involved in the Cambridge trial partners were Adaptrum, Alcatel-Lucent, Arqiva, BBC, BSkyB, BT, Cambridge Consultants, CSR, Digital TV Group, Neul, Nokia, Samsung, Spectrum Bridge, TTP, Virgin Media, and Microsoft.

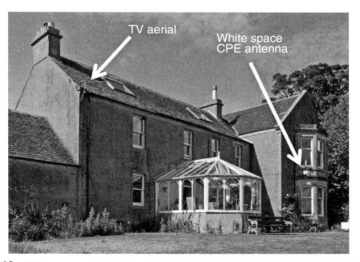

Figure 5.12   White space and TV antennas at one of the installation sites in Isle of Bute. (Reproduced from Ref. [6] with permission from Centre for White Space Communications.)

5.2.1.3 Framework Testing

The third part of the pilot is to test that whether the TV white space framework is fit for providing broadband access to remote rural areas and also for understanding what revisions might be required or are desirable for the full solution framework. The framework testing includes two parts: business process testing and end-to-end testing.

The business process testing mainly focused on the mechanisms and processes that Ofcom used to communicate with WSDBs during the pilot. The objective of the testing is to verify that these mechanisms and processes were fit for purpose, and also to understand whether any amendments or improvements may be required for the operational solution.

There were two issues emerged during the business process testing. The first issue is that each of the WSDBs implemented a white space devices information system(WSDIS). This means if interference occurs, the interference management team at Ofcom would have to carry out multiple queries across multiple systems. This is very time-consuming and makes the triage of possible WSD interference very complicated. Ofcom is currently working with WSDBs to simplify the process. The second issue is that the "cease transmissions" function is not working properly after the function is conducted several times. The reliability of this cease transmission function is the important issue Ofcom needs to address.

5.3 ASIA

Table 5.4 summarizes the TV white space trials in Asia, while the detailed descriptions are provided in the following sections.

5.3.1 Bhutan

Of the Bhutan's total population, 80% live in areas where health care facilities are usually more than an hour's walking distance. It can take an ambulance more than 20 h to reach the capital Thimphu, where there are more extensive health facilities and services. Tang is one of the areas where people desire more health facilities.

In fact, Bhutan has successfully tapped e-Health to meet the demand for quality and accessible health care. It involves the Internet and other related tele-communication technologies to facilitate efficient communication and delivery of health services. However, Himalayan Mountains, valleys, and subtropical forests deteriorate terrestrial broadband connectivity. Under such circumstances, the cost of deploying conventional broadband solutions such as cellular, terrestrial, and satellite-based connections is very high and will not be able to intrigue investors' interests.

The establishment of the Internet infrastructure in Tang is obstructed because of its mountainous terrain. However, now TV white space provides an excellent chance to improve the Internet infrastructure because of the spectrum characteristic of TV white space(can transmit longer and penetrate into materials easier). With the TV white space technology, point-to-multipoint broadband connectivity over long

Table 5.4 TV White Space Trials in Asia

Service Provider	Devices	Trial Location	Trial Application
Microsoft, Terabit Inc., Asian Development Bank	Terabit Inc.	Tang, Bhutan	Health care
Microsoft, DOST-ICT, DA-BFAR, USAID	Power Automation	Tubigon, Bohol, Philippine	Rural broadband
DOST-ICT	Power Automation	Bohol/Visayas, Philippine	Disaster communications
Ministry of Internal Affairs and Communications	NICT	Tono City, Japan	Disaster communications
SWSPG	Power Automation	Fu-Hsing township, Taoyuan County, Taiwan	Wireless broadband services
SWSPG, IDA	Power Automation	Singapore	Wireless broadband, external Wi-Fi backhaul, video surveillance
Indonesia Ministry of Communications and Information Technology (MCIT), Microsoft	NICT	Indonesia	Rural broadband
Hong Kong Science & Technology Park	I²R	Hong Kong	Video surveillance

distances can be achieved at low cost. The broadband connectivity brought by TV white space can let people book appointments and receive medical advice without the long journey into the city.

Microsoft, Terabit (Bhutan Telecom), and the Asian Development Bank (ADB) cooperate with each other to connect a remote health unit in the Bhutanese village of Tang to Bumthang Hospital. In February 2014, they established the TV white space link that connects the villagers to quality health care and ensures that treatment can be received quickly and cost-effectively (Figure 5.13).

The pilot is being used by ADB for designing an ICT development loan instrument to leverage the cost-effectiveness of TV white space technology. The overall architecture is shown in Figure 5.14.

5.3.2 The Philippines

Microsoft, the Philippine Department of Science and Technology's Information and Communication Technology Office (DOST-ICT Office), Department of Agriculture's Bureau of Fisheries and Aquatic Resources (DA-BFAR), and the U.S. Embassy Manila's the United States Agency for International Development

Figure 5.13 Before and after (*inset*) the installation of the TV white space antenna and base station at a health center in Tang, Bhutan (*Photo:* By TA 8418 implementation team/ADB). (*Source:* http://www .adb.org/sites/default/files/publication/42599/developing-ehealth-capabilities-bhutan.pdf [7]. Reproduced with permission of ADB.)

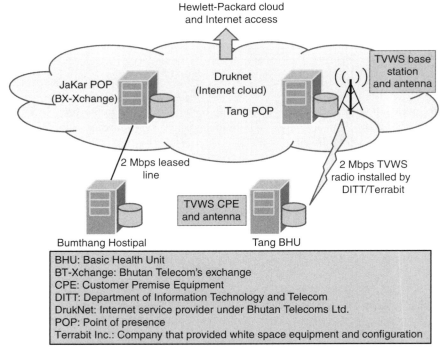

Figure 5.14 Television white space antenna and base station system configuration at a health center in Tang, Bhutan. (*Source:* http://www.adb.org/sites/default/files/publication/42599/developing-ehealth-capabilities-bhutan.pdf [7]. Reproduced with permission of ADB.)

Figure 5.15 Internet connectivity testing at Tubigon through Wi-Fi via TV white space. (Courtesy of Kerk See Gim, Power Automation.)

(USAID) have a partnership to explore providing wireless broadband access and facilitating mobile fishing registration to remote areas in the municipalities of Talibon, Trinidad, Bien Unido, Ubay, and Carlos P Garcia.

A TV white space base station was deployed at a transmission tower at a compound in Tubigon, Bohol to bring Internet access to the population all around the island. Currently the residents there only have slow or even no network connectivity. The base station is set up in a way that the main communication can cover the 3–7 km radius surrounding areas of the base station. The TV white space receiving clients are deployed at local schools or community areas within the villages (Figure 5.15).

The trial also allowed the fishermen in remote locations to obtain the IDs, certificates, and licenses from local government agencies collating the online database. It also allowed the police to access the central database to monitor compliance and cross reference individuals. It can be expected that with the enhanced support from the government, the village will be able to produce more fish, feed more people, and generate more jobs. For further expansion of the TV white space system, the pilot will also provide wireless broadband access to local institutions such as clinics and public schools (Figure 5.16).

On the other hand, on November 8, 2013, Typhoon Haiyan, one of the strongest typhoons ever recorded, struck the Philippines and killed at least 6268 people and affected approximately 11 million more. The typhoon also caused the whole communication network failure and infrastructure damage, and this significantly added to the difficulty in the rescuing and relieving mission.

The DOST-ICT office built a TV white spaces network to provide an immediate on-the-ground communications system for disaster relief respondents and victims to use. The built TV white space system provided a network immediately for two-way voice and data wireless communications for anyone carrying a functioning

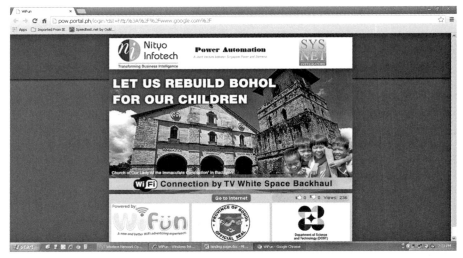

Figure 5.16 The portal Web site for the wireless Internet trial provided by power automation. (Courtesy of Kerk See Gim, Power Automation.)

end user device (handsets, laptops, tablets, etc.). The cost of building the TV white space network is much lower than (around 1/10th the cost) the other comparable solutions. It can be set up in a very short time, and requires only little technical expertise, system integration, or support.

The Philippines pilot tests for white space broadband are the most extensive in Asia. A series of pilot tests started after the Bohol earthquake and Typhoon Haiyan in 2013. They helped to provide critical disaster relief and allowed communications throughout the affected areas of Bohol and Leyte provinces in the Visayas. It is also believed that the Philippines is an ideal location to test white space radio capability, because the country is featured with its thick foliage, remote locations, and widespread villages.

5.3.3 Japan

NICT and Hitachi Kokusai Electric carried out a project under R&D contracts with the Ministry of Internal Affairs and Communications, Japan. The project is to establish a long-distance broadband communication system using TV white space technology for disaster scenarios. The TV white space technologies used in this project conducted in Tono City are IEEE 802.22 and IEEE 802.11af. There is a key assumption of the pilot that TV white space system would provide Internet access where commercial network services such as cellular networks were damaged by an earthquake. In the project, the IEEE 802.22 system established a link over 12.7 km, with the throughput equal to 5.2 Mbps downstream and 4.5 Mbps upstream. NICT has modified the original specifications of IEEE 802.22 so that it can operate on multiple noncontiguous channels with channel aggregation technologies. And if two channels were used, throughput can be up to 15.5 Mbps downstream and 9.0 Mbps upstream over 6.3 km (Figures 5.17 and 5.18).

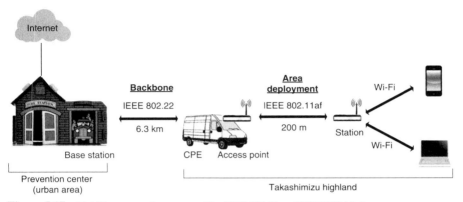

Figure 5.17 Multi-hop network constructed by IEEE 802.22 and IEEE 802.11af.

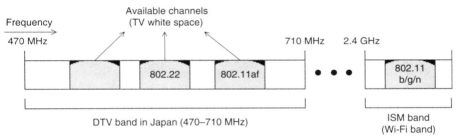

Figure 5.18 Illustration of channel usage in Tono City's trial.

In this project a multi-hop network constituted by IEEE 802.22 and IEEE 802.11af systems was built. The IEEE 802.22 provided for backhaul lines while IEEE 802.11af systems provided for expansion of service area. Conventional Wi-Fi network was provided for the last 100 m and was converted to the IEEE 802.11af system so that end users using Wi-Fi devices can connect to the Internet through the TV white space technologies.

5.3.4 Taiwan

Power Automation, Institute for Infocomm Research (I^2R, Singapore), and local research partners conducted a TV white space pilot in Fu-Hsing township, Taoyuan County, Taiwan [8]. Fu-Hsing township is an aboriginal rural township with eco-system tourism spots. The mountainous terrain of this area makes deployment of broadband Internet access challenging. The pilot project built a TV white space system covering an area of $350 \, km^2$ from a single base station that can connect to remote sites 10–20 km far away. The remote site is then connected to the 2.4 GHz Wi-Fi hotspot for public access. The trial project is meant to lay the foundation for deploying nationwide free Wi-Fi access for residence and tourists, supported by the

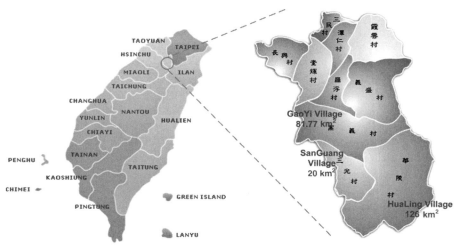

Figure 5.19   TV white space trial locations in Taiwan.

government as part of the iTaiwan initiative. With such a deployment, mountainous location can benefit from TV white space to support mobile medical vehicles and promote trade and tourism (Figure 5.19).

5.3.5 Singapore

The first wave of the TV white space pilots led by Singapore White Spaces Pilot Group (SWSPG) was announced in September 2012. These pilots aim to establish TVWS projects that demonstrate the variety of commercial services that could be deployed using white spaces technology in an environment where traditional wireless system could be difficult to be deployed. SWSPG was established in April 2012 with four founding members (I^2R, Microsoft, StarHub, and Neul) and was supported by Infocomm Development Authority (IDA, Singapore). The organization's objective is to deploy white spaces technology pilots in Singapore, thereby expediting the adoption of TVWS technologies in Singapore, the Asia Pacific region, and globally. Under the SWSPG umbrella, various commercial pilot trials have been carried out in Singapore. Some of the trials have been completed and others are in progress.

A TVWS concentrator was placed at level 25 of a University Town (UTown) South Tower. TVWS client nodes were placed at the rooftop of the pump house ($\sim$0.40 km distance to tower), building SR1 at ground level ($\sim$1.26 km distance to tower), and building SR2 at the water tank level ($\sim$1.04 km distance to tower). Data from SR1 and SR2 were transmitted to the base station on UTown South Tower, which directed the data to the Pump House. The measurement results of this pilot are shown in Table 5.5.

In the Singapore Island Country Club pilot, the distance between the node locations to the base station ranges from 376 m (Pole) to 1.24 km (Workshop), as

Table 5.5   Measurement Results for the Smart Metering Trial in NUS, Singapore

Client	Antenna (dBi)	Channel	Max. Speed (Mbps)	Min. Speed (Mbps)	Link Quality	Estimated Distance (Km)
SR1 above water tank	12	48	2.67	1.82	19	1.26
SR1 ground level	12	48	2.95	1.33	8	1.26
SR2 above water tank	12	48	4.46	4.05	25	1.04
SR2 ground level	12	48	1.85	1.07	13	1.04
Pump house above water tank	12	48	6.53	6.30	58	0.40

shown in Figure 5.20. The terrain has different elevation, with up to 16–36 m difference in elevation level between base station and the other nodes.

Another TV white space pilot was targeting at the residential areas in Singapore. The communications platform will enable collecting different information to enhance security and safety in residential areas and provide various users a cost-effective method for local connectivity to field devices installed in various corners of the estate. Figure 5.21 shows the coverage of Housing Development Board (HDB) blocks with each oval representing a TVWS coverage zone.

The Sentosa Island trial aimed to set up a TV white space network at Sentosa Island to connect Siloso beach with Merlion. However, the trial revealed that there is currently no clear frequency band available in the 630–742 MHz range for stable operation.

Figure 5.20   Singapore Island Country Club Course as on April 12, 2014, Terrabit networks. (*Source:* SDR-WInnComm Annual Report, Section 6. Reproduced with permission of WinnForum.)

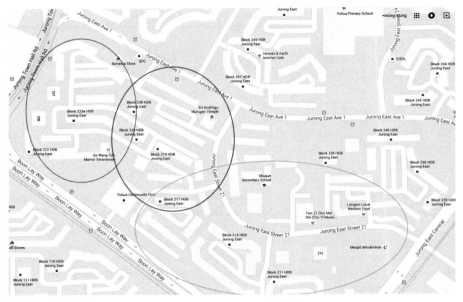

Figure 5.21 Coverage area of the HDB TV white space trial in Singapore.

The park management of Gardens by the Bay also initiated a TV white space project to provide a cost-effective solution to provide free wireless Internet access for public users around the main attractions, as well as to provide video surveillance at locations where regular events are organized. TV white space technology deployed in the corners of the attractions together with the free public Wi-Fi implemented by Wireless@SG already provided a scalable and cost-effective Wi-Fi access solution for local and overseas visitors to enjoy Internet connectivity at the Gardens. The overall deployment at Gardens by the Bay is shown in Figure 5.22.

In addition to the activities hosted by SWSPG, StarHub and InterDigital also conducted an evaluation and trial of Wi-Fi over dynamically managed unlicensed spectrum. The DSM Wi-Fi platforms were used to evaluate the suitability of Wi-Fi and TV white space spectrum for both indoor and outdoor applications. The key applications running on this platform include video surveillance and small cell. Testing (on throughput, latency, and coverage) showed good performance with minimal latency and TCP/IP speeds up to 20 Mbps over a 40 m indoor NLOS distance.

Recently, an I-Man Facility Sprinter (IFS) pilot run in Singapore also utilized the TVWS technology as its communications system. IFS is a custom-built vehicle equipped with security and building monitoring systems, including sufficient DVRs, CCTV monitors and other building monitoring equipment. IFS can be used to replace the physical deployment of manpower to the building premises, and is able to monitor fire alarm panels, lift control panels, switch rooms, sprinkler systems, pump rooms, water tank compartments, generator rooms, car-park barrier systems, intercom systems, pressurization and exhaust fan control panels, hose reel

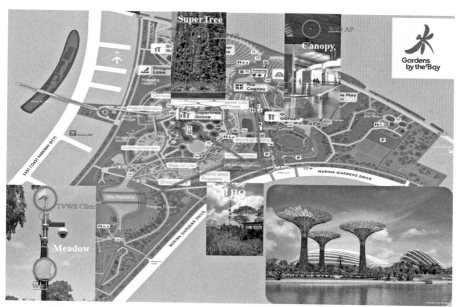

Figure 5.22   Deployment of Wi-Fi access points and surveillance cameras using TV white space technology at the Gardens by the Bay, Singapore. (*Source:* SDR-WInnComm Annual Report, Section 6. Reproduced with permission of WinnForum.)

tanks, cooling towers, air-conditioning systems, and so on. With these functions provided, IFS is able to provide security protection and building services for a cluster of building premises. Multiple IFSs deployment will form a mesh network of high efficiency and connectivity by utilizing the TVWS technology.

The IFS utilizes TVWS network to deliver various kinds of business services. By using the range and penetration advantages, coupled with potentially abundant bandwidth, TVWS offers IFS the network transmission efficiency, which current 3G and Wi-Fi service providers are finding difficult to offer (Figure 5.23).

5.3.6 Indonesia

Despite the rapid economic growth in Indonesia in recent years, the Internet connectivity in Indonesia is still behind most of its neighboring countries. This problem is even more severe in the rural areas. To mitigate this problem, the Government of Indonesia recently launched Indonesia Broadband Plan (IBP, 2015–2019), which targets on investigating various technologies that could provide wireless broadband connectivity to rural areas.

In this IBP plan, the Indonesia Ministry of Communications and Information Technology (MCIT) teamed up with Microsoft, Hitachi, and the Government of Japan to conduct a pilot leveraging on the IEEE 802.22 technology to see how feasible the TV bands can be leveraged for improving the Internet penetration in rural areas.

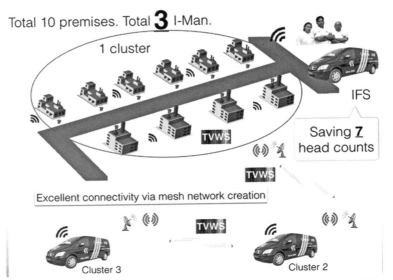

Figure 5.23 IFS deployment and scenario in Singapore. (Reproduced with permission of Concorde Security.)

The pilot took place in the Yogyakarta region of Indonesia between early 2014 and March of 2015. There were two wireless broadband links established in the pilot, and these links are utilized by a school, a health clinic, a public Internet center, and a small agribusiness. The radio link performance was tested by the pilot evaluation team, and the results were used to assess whether TVWS would be cost-effective or would have social and economic impact.

The results of the pilot indicate that it is both technologically and financially feasible to meet IBP's targets using the TVWS technology. Both links provided sufficient bandwidth (8.5 and 9 Mbps, respectively) for use as distribution local access networks in rural areas. Based on the cost of establishing the TVWS link, the pilot also demonstrated a strong economic impact through providing a positive financial return to the network operator. Similarly, the pilot also indicated that MCIT's investments in subsidizing rural connectivity through the Indonesian Universal Service Fund resulted in a good return. The pilot also facilitated the ongoing regulatory discussion involving the highest levels of government, and thus provided potential regulatory options that are still under consideration.

5.3.7 Hong Kong

In Hong Kong, Microsoft teamed up with researchers in Chinese University of Hong Kong to run a project that focused on exploring the TVWS indoor usage. The team first confirmed that there are abundant white spaces available. In particular, more than 50 and 70% of the TV spectrum are white spaces in outdoor and indoor scenarios, respectively. To further utilize the indoor TVWS spectrum, the team proposed the White-space Indoor Spectrum Enhancer (WISER) to identify and track indoor white spaces in a building, without requiring user devices to sense the spectrum. In this project, a WISER prototype is built and real-world experiments are conducted to evaluate the performance of WISER. Their results showed that WISER can identify 30–50% more indoor white space with very few false alarms presented [9], as compared to some baseline approaches.

After the successful IFS pilot, Concorde Security, a Singapore based company, teamed up with I^2R to deploy the same IFS pilot in the Hong Kong Science & Technology Park. They installed a TVWS base station at the highest floor of a commercial building (as shown in Figure 5.24) and a TVWS client in a moving vehicle equipped with security and building monitoring systems. It has been shown in the pilot that the TVWS link is able to carry the video surveillance data stream even in a long-distance test case with the vehicle moving constantly.

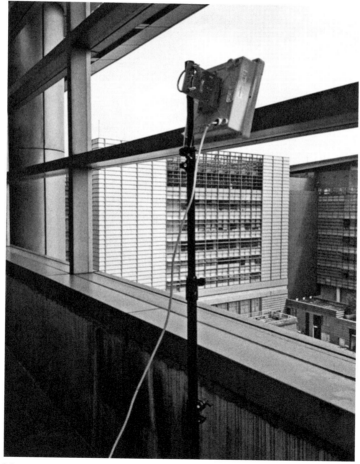

Figure 5.24 The installation of TVWS base station facing down to the road for the Hong Kong IFS pilot. (Reproduced with permission of Concorde Security.)

5.4 AFRICA

Microsoft 4Afrika initiative,[2] the key proponent of the TV white space develop-ment in Africa, is a multi-year initiative that represents Microsoft's increased com-mitment to Africa as it celebrates its 20 years of doing business on the continent. By looking forward to the next 20 years, Microsoft wishes to explore new ways to link the growth of business with initiatives that accelerate growth for the continent by focusing on three critical areas: world-class skills, access, and innovation.

The goal of Microsoft 4Afrika initiative is to empower people's ability to turn their ideas into reality, which in turn can help their community, their country, or even the continent at large. The Microsoft 4Afrika Initiative is built on the mutual belief that technology can accelerate growth for Africa, and Africa can also contrib-ute technology to the world.

By viewing Africa's TV white space trials listed in Table 5.6, we can see most of the trials are connected to the Microsoft 4Afrika initiative. The details of these trials are described in the following sections.

Table 5.6 TV White Space Trials in Africa

Service Provider	Devices	Trial Location	Trial Application
Microsoft, Botswana Innovation Hub	Adaptrum	Botswana	Health care
Microsoft, MyDigitalBridge Foundation	Adaptrum	Oshana/Ohangwena/ Omusati, Namibian	Wireless broadband services
SpectraLink Wireless	6Harmonics	Meltwater Entrepreneurial School of Technology, Accra, Ghana	Wireless broadband services
Microsoft, Facebook, SpectraLink Wireless	6Harmonics	Koforidua, Ghana	Wireless backhaul
Tertiary Education and Research Network of South Africa	Carlson Wireless	Cape Town, South Africa	Research and development
Microsoft, Meraka Institute	6Harmonics	University of Limpopo, Polokwane, South Africa	Education services, wireless backhaul
Malawi Communications Regulatory Authority	Carlson Wireless	Zomba, Malawi	Wireless backhaul
UhuruOne, Microsoft	Adaptrum	Dar es Salaam, Tanzania	Wireless broadband services
Microsoft, Mawingu Networks, Jamii Telecom Limited	Adaptrum, 6Harmonics	Nanyuki, Kenya	Wireless broadband services

---

[2] http://www.microsoft.com/africa/4afrika/

5.4.1 Botswana

The Botswana Innovation Hub, in collaboration with Microsoft, the Botswana–UPenn Partnership (BUP), Global Broadband Solutions, Vista Life Sciences, BoFi-Net, Adaptrum, and USAID-NetHope, launched a TV white space pilot project named Project Kgolagano. The goal of the project is to provide Internet connectivity and telemedicine services to local hospitals and clinics, which will facilitate a remote diagnosis network and support local health care service delivery.

Project Kgolagano, which means 'to be connected or networked', has a specific focus on providing access to specialized maternal medicine, which will enhance the health condition of women located in small towns and rural areas. In this project, telemedicine experts and doctors providing medical expertise for referred patients are being provided by Botswana–UPenn Partnership.

Project Kgolagano has obtained an authorized permission from the Botswana Communications Regulatory Authority (BOCRA) to transmit using TV white spaces. The project was also conducted with the support from the Botswana Ministry of Health and the Ministry of Infrastructure, Science and Technology.

5.4.2 Namibia

MyDigitalBridge Foundation in collaboration with Microsoft and Adaptrum, with support from the Millennium Challenge Corporation (MCC) and Millennium Challenge Account (MCA)-Namibia, has successfully trialed the Namibian TV white spaces pilot project. The intention of the project is to provide a blueprint of broadband Internet connectivity countrywide. The pilot consists of a network deployed over a $62\,km \times 152\,km$ ($9424\,km^2$) area covering three regional councils—Oshana, Ohangwena, and Omusati—and connecting 28 schools in northern Namibia. This is already the biggest TVWS project of its kind in terms of area coverage.

At the University of Namibia's Jose Eduardo dos Santos Campus in Ongwadiva, Namibia, the trial was demonstrated to the Namibian Parliamentary Standing Committee with the following results. The TV white space network connected three regional offices, 28 schools, and seven education circuit offices, all with a link distance of 8–10 km (with two links at 12 km). Typical speeds range from 5 to 10 Mbps with the help of the Adaptrum ARCS 2.0 TVWS radios, and the network provided users with a wide range of voice, video, and data applications. The demo included high-resolution Skype video conferencing from three locations, each connected to the Internet with the help of Adaptrum's TV white space technology.

5.4.3 Ghana

In March 2014, SpectraLink Wireless, under authorization from the Government of Ghana's National Communications Authority (NCA), cooperated with the Meltwater Entrepreneurial School of Technology (MEST) to deploy a pilot network to offer free wireless broadband access for its community of entrepreneurs in training.

The pilot that successfully established the leading edge TV white space network is the first of its kind in West Africa. It was also proven in the pilot that SpectraLink's platform in delivering high-speed Internet connectivity through TV bands is viable. The purpose of the pilot with MEST has been to test the efficiency of using TV white spaces in an urban environment where there are multiple sources of interference. It has been shown in the pilot that there was no obvious interference observed on the link (10 km long) while the channels adjacent to active television channels were used.

Another pilot in Ghana is the collaboration result of SpectraLink Wireless, Microsoft, and Facebook. The pilot connects Koforidua Polytechnic and other universities in Ghana to wireless broadband and cloud-based services through TV white space technology. The pilot is supported by Microsoft's 4Afrika initiative, which aims to improve the region's competitive positioning by facilitating ICT technologies and ensuring students have reliable Internet access.

The networks use TV white space radio devices from 6Harmonics to connect campus buildings and hostels where students live, to ensure pupils to have fast and free wireless broadband access and cloud-based services such as Office 365. The operation of the first TV white space pilot project was licensed by the Ghana National Communications Authority.

The TV white space radio characteristics enable greater Internet penetration, providing more bandwidth at a lower cost, and allow students to take university courses online and use bandwidth-consuming applications such as video conferencing.

5.4.4 South Africa

In 2011, several organizations including the Independent Communications Authority of South Africa (ICASA) hosted a workshop in Johannesburg about the TV white space technology. After that ICASA gave support for a white space trial in South Africa, which is conducted by the group involving Google, Meraka Institute, Tertiary Education and Research Network of South Africa, e-Schools Network, the Wireless Access Providers' Association, Comsol Wireless Solutions, Carlson Wireless, and Neul. The trial was launched in Cape Town in March 2013 and was continued for 6 months. The TVWS network consisted of multiple base stations located at Stellenbosch University's Faculty of Medicine and Health Sciences in Tygerburg, Cape Town and delivered broadband Internet services to 10 schools within 10 km radius area. Each school can have a dedicated 2.5 Mbps bandwidth and an alternative ADSL link so that the school can still have connectivity during the down time of the TV white space network. To prevent interference with other systems, the network used Google's spectrum database to determine which channel to use. At the end of the 6-month period, the trial has successfully shown that broadband can be offered through TV white spaces without interfering with licensed spectrum holders.

In 2013, the University of Limpopo in South Africa partnered with Microsoft, 6Harmonics, and The Meraka Institute, which developed white space databases, to provide the wireless Internet access to an institution and five schools within the 8 km radius area of the university. The main base station is on the University of Limpopo campus and is powered by solar energy. As a result, hundreds of students are able to benefit from TV white space technologies that allow ICT to be included in the syllabus for the very first time. The project enables students to access vast online resources and very-high-quality education that can improve the employment opportunities for these students.

The cost of broadband is very high so that Internet access is still an obstacle to rural South Africa. By reducing the cost of broadband access, millions of South Africans could go online more easily. Doing so will also create new opportunities for education, health care, commerce, and the delivery of government services across the country.

It is hoped that the TV white space pilot could be extended to provide Internet access to disadvantaged schools throughout South Africa and deliver interactive-based education to more students in the country. The focus of this pilot is to prove that TV white space technology can be used to meet the South African Government's goals of providing low-cost access for a majority of South Africans by 2020.

Google was the first major player in the country, and has been deploying a similar project for 10 local schools in Cape Town in 2013. The town is surrounded by hills and mountains; thus, the TV white space is an appropriate solution to such a condition. The main base station is connected to a 10 Gbps backhaul.

5.4.5 Malawi

The University of Malawi, in partnership with the Malawi Communications Regulatory Authority (MACRA), and the International Centre for Theoretical Physics in Trieste, has launched a TV white space pilot project in the city of Zomba, in southern Malawi. The pilot, which embarked from September 2013, has connected a number of institutions, including a school, a hospital, an airport, and a research facility. A performance improvement over other broadband access technologies has been measured on link distances of up to 7.5 km.

5.4.6 Tanzania

UhuruOne and Microsoft are partnering with the Tanzania Commission for Science and Technology (COSTECH) to launch a project called Broadband4Wote, by aiming to provide affordable wireless broadband access, devices, and cloud-based services to 74,000 students in Tanzania.

The Broadband4Wote project is designed to assess the commercial feasibility of wireless broadband using the TV white space technology. The Open University of Tanzania, Institute of Finance Management, College of Business Education, and Dar Es Salaam School of Journalism and Mass Communication are all connected to the network, and then the capability of the TV white space technology applied in the urban environment is also assessed. The ongoing pilot also hopes to improve

student's education level through access to vast online resources and boost employment opportunities through the development of the new wireless technology.

5.4.7 Kenya

Microsoft, in collaboration with the Government of Kenya (Ministry of Information
and Communications), Mawingu Networks (previously Indigo Telecom), and Jamii
Telecom Limited, is conducting a pilot project under the support of Communications Authority of Kenya (CAK) that can deliver low-cost wireless broadband
access to underserved schools, health care clinics, government offices, and some
small businesses near Nanyuki, Kenya. The network launched in 2013 utilizes
Adaptrum and 6Harmonics TV white space and solar-powered base stations to
deliver broadband access and create new ways for commerce, education, health
care, and delivery of government services.

   To maximize coverage and bandwidth, and keep the cost minimized, the project is using several complementary spectrum bands available to license–exempt end
devices. These spectrum bands include 13, 5, and 2.4 GHz bands as well as TV
white space band. The Mawingu project has successfully demonstrated the technical viability of this model of delivery, with point to multipoint coverage of up to
14 km from the TV white space base station operating at only 2.5 W power (EIRP
measurement) without causing any interference. It has been shown that the TV
white space technology is able to communicate at a data rate of up to 16 Mbps on a
single 8 MHz TV channel at a distance of up to 14 km.

5.5 THE REST OF THE WORLD

5.5.1 Uruguay

In Latin America, Microsoft is involved in the region's first white spaces pilot in
Uruguay, in which the company is providing technical support to Plan Ceibal, an
initiative integrating information and communications technologies with the country's public education system. The TV white space technology used in Plan Ceibal
is provided by 6Harmonics, a member of Dynamic Spectrum Alliance, with the
capability of providing broadband access to 10 rural schools.

5.5.2 New Zealand

The managed spectrum park is a brand new experiment by New Zealand's regulator, which is based on the concept that the spectrum allocation and the use of spectrum should be handled by a self-managed approach. In each region of the park, the
entry is on a first-come first-served basis.

   The guideline to the park licensees is that they have to resolve themselves for
the spectrum allocation and interference issues, and they also need to cooperate

with each other while resolving these issues. If it happens that demand is more than supply, then ballots are drawn to reduce the number of applicants in a region.

The regulator is closely monitoring how the experiment is going in the managed spectrum park. Until now, allocations are reported to have been "somewhat contentious as applicants have found that the length of the application process and possibility of challenges to be frustrating" [10].

REFERENCES

1. H. Nwana, Spectrum sharing omens bode well. EETimes Blog, 2014.
2. Wireless Innovation Forum, Dynamic Spectrum Sharing Annual Report: 2014, Document WINNF-14-P-0001, August 2014.
3. Broadband Center of Excellence, TV white space: ready for prime time? Assessing practical realities of a share-spectrum approach for broadband Internet access, Broadband Intelligence Series, University of New Hampshire, January 2014.
4. John Gavan, Super Wi-Fi (TV whitespace) networking & extending library services into the community, Delta County Library District, June 2014.
5. Ofcom, Implementing TV White Spaces, February 2015.
6. Centre for White Space Communications, Final Report: White Space Rural Broadband Trial on the Isle of Bute,TSB100912, June 2013.
7. Asian Development Bank, Developing e-health capabilities in Bhutan, Knowledge Showcases, Issue 57, June 2014.
8. Power Automation, Proof of Concept (POC) Test Report for TV Whitespace (TVWS) in Fu-Hsing Township, September 2013.
9. X. Ying, J. Zhang, L. Yan, G. Zhang, M. Chen, and R. Chandra, Exploring indoor white spaces in metropolises, in Proceedings of ACM MobiCom, Miami, Fl, September 30–October 4, 2013.
10. New Zealand Ministry of Business, Innovation and Employment, Radio Spectrum Five Year Outlook 2012–2016, July 2013.

Chapter 6

Commercial and Market Potential

6.1 INTRODUCTION

According to the survey done by Richard Thanki [1], for over 3.9 billion people or around 61% of the world's population, the price of fixed broadband is still unaffordable. The difference is also huge between different continents: 8% of the population in the Europe cannot afford fixed broadband and 90% of the population in Africa cannot afford it. Similarly, basic mobile broadband service is also unaffordable for over 2.6 billion of the world's population.

Moreover, the coverage of the broadband network is far from omnipresent. There are fewer than 1.2 billion telephone lines serving the world, and only a small proportion of the telephone lines are able to deliver broadband service. Mobile broadband covers only around 35% of the world's population. Main reasons contributing to world's broadband availability and affordability gap are the high costs and high barriers to enter the world's fixed and mobile telecom industries. From Figure 6.1 we can see that telecom operators are targeting merely at the richest 3 billion people in the world because of the high cost of building a mobile telecom system. To support more users telecom operators have to expand their current cellular system with additional investment, and that additional cost will be disproportional to the increase in return. The diminishing return through expansion of mobile broadband network to less populated areas can be mainly attributed to two reasons. First, the nearest fiber is likely to be some distance away, thus, the cost of laying fiber to the additional base station is higher. Second, even if a new base station is set up, the mobile population that it covers is lower compared to densely populated areas. Thus, the average cost per mobile user increases drastically as shown on the left-hand side of Figure 6.1. Moreover, service providers have been changing their business models that rely more and more on mobile networks (e.g., Google started offering mobile service, Microsoft teamed up with Spectral Link Wireless to offer in-building wireless communication and services), telecom operators really need to increase their system capacity by all means.

TV White Space: The First Step Towards Better Utilization of Frequency Spectrum, First Edition.
Ser Wah Oh, Yugang Ma, Ming-Hung Tao, and Edward Peh.

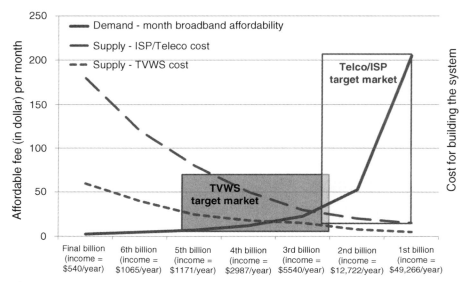

Figure 6.1   Broadband affordability for different income groups of the world. (Data from Ref. [2].)

On the other hand, technologies that use license–exempt spectrum are cost-effective and can be deployed by any person or entity. The TVWS technology is a good example, which is affordable even for the people with middle/low income, as shown in Figure 6.1. TVWS system can accommodate more users (even to the 5th billion group) with the cost that is much less than that in the cellular system. In fact, these technologies have been deployed by existing operators, new entrants, and individuals who wish to increase the quality, decrease the cost, and extend the reach of broadband networks.

The comparison in Figure 6.1 does not take into consideration the quality of the communication link. Some may argue that broadband over fiber or 3G/4G with licensed spectrum has better data rates or quality of service (QoS) compared to TVWS. This observation may be true for now, but may not hold in the future since TVWS is a disruptive technology as depicted in Figure 6.2. At this point in time, TVWS broadband might be acceptable for developing countries or people with lower broadband requirements. In a disruptive curve, technologies typically advance much faster than user expectations. When this trend goes on, TVWS will be able to fulfill the needs of most broadband users, significantly complementing the services provided by traditional telecom operators or service providers. Therefore, operators should start looking into using TVWS as a complementary system to their current cellular networks.

To largely increase the capacity, telecom operators are pushing their LTE system to utilize the license–exempt spectrum (5 GHz) as well. Currently there are two ways that allow the usage of 5 GHz spectrum in the LTE system LTE-U and LAA-LTE. LTE-U is the version of LTE unlicensed that was proposed in 2013 by

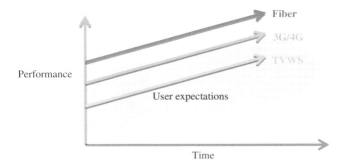

Figure 6.2 TVWS following Clayton M. Christensen's Disruptive Technology curve.

Qualcomm and Ericsson. LTE-U is based on the functionality provided by 3GPP Release 10–12, with specifications defined by the LTE-U Forum. LTE-U forum is an organization formed by Verizon, Alcatel-Lucent, Ericsson, Qualcomm Technologies Inc., and Samsung. Because LTE-U requires lesser modifications from the original licensed LTE, it will be the first version of unlicensed LTE to be available in commercial deployments. However, because it does not implement listen-before-talk mechanisms, LTE-U can only be used in countries where regulation does not require LBT, such as China, Korea, India, and the United States.

LAA-LTE is another version of unlicensed LTE that 3GPP plans to standardize in Release 13. Unlike LTE-U, LAA-LTE firmly supports listen-before-talk mechanism. The standardization efforts that LAA-LTE has taken so far are to ensure that it can meet regulatory requirements worldwide. However, because the standardization work is not yet completed, commercialization will take longer time than that in LTE-U. In the long term, it is expected that operators and vendors worldwide are going to support LAA-LTE because it provides a globally harmonized solution that leads to better interoperability among equipment and devices.

Wi-Fi, the representative license–exempt technology has already dominated world's data traffic, as shown in Figure 6.3. In the case of smartphones and tablets,

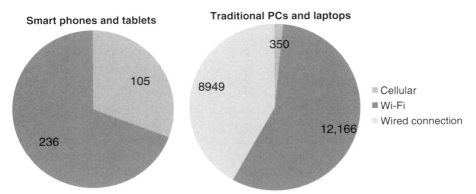

Figure 6.3 Traffic carried by different networks for different types of devices (petabyte per month). (Data from Ref. [1].)

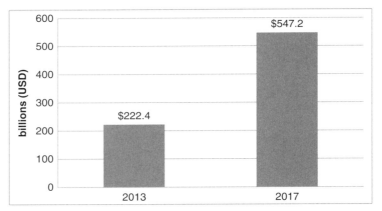

Figure 6.4   Current and future economic value of unlicensed spectrum in the United States. (*Source*: Data from Ref. [3].)

Wi-Fi carries 69% of the total traffic generated. For traditional PCs and laptops, Wi-Fi carries 57% of total traffic, which is more than the share of wired connection (e.g., ADSL, Cable/Fiber broadband) and cellular data combined.

In the United States, the combined value of the license–exempt spectrum and its application by 2017 is estimated to be at least $547.22 billion in economic surplus (Figure 6.4) annually and $49.78 billion in contribution to the annual GDP (Table 6.1) [3].

Table 6.1   The United States: Summary of Future Economic Value of Unlicensed Spectrum (2013–2017)

Drivers	Technologies and Applications	Economic Surplus		GDP Contribution	
		2013 ($)	2017 ($)	2013 ($)	2017 ($)
Future values of currently deployed technologies and applications	Wi-Fi cellular off-loading	12.60	22.60	3.102	7.033
	Residential Wi-Fi	36.08	268.74	NA	NA
	Wireless internet service providers	NA	NA	1.439	4.80
	Wi-Fi only tablets	42.87	47.99	NA	NA
	Wireless personal area networks	NA	NA	2.166	1.652
	RFID	130.87	191.46	NA	NA
	Subtotal	222.38	531.02	6.778	13.485
Value of not yet adopted technologies and applications	High speed wireless	—	NA	—	4.81
	M2M	—	NA	—	31.49
	Smart city deployments	—	15.10	—	0.79
	Agriculture automation	—	1.10	—	NA
	Subtotal	—	16.20	—	36.30
Total		222.38	547.22	6.778	49.78

In US billion dollars.
Source: TAS analysis.

Table 6.2 Economic Value Generated by Wi-Fi Through Fixed Broadband Value Enhancement [1]

Continent	Low Value (Million Dollars Per Year)	High Value (Million Dollars Per Year)	Wi-Fi Connections (Million)
Africa	69	901	0.5–1
Asia	10,820	41,516	21.2–48.2
Europe	21,657	30,164	15.4–35
North America	17,769	19,952	10.2–23.2
Oceania	1049	1217	0.6–1.4
South America	782	4772	2.4–5.5

Due to the presence of Wi-Fi and the corresponding unlicensed spectrum, fixed broadband service has increased its value significantly. It is because Wi-Fi can let a broadband connection shared by multiple individuals easily. A total of 439 million households (around 25% of all households worldwide) have home Wi-Fi networks. Each household may derive a yearly benefit of $118–$225 from Wi-Fi, which results in a total economic gain for all households of around $52–$99 billion annually. The breakdown of the economic values generated by Wi-Fi by continents is listed below in Table 6.2. If there is no Wi-Fi, the value of fixed broadband would not be as high as now and perhaps the number of fixed broadband connections will decrease up to 114 million around the world.

Wi-Fi also helps to decrease the costs of cellular data networks in multiple ways (Wi-Fi off-loading is one of them). It is also found from recent research that Wi-Fi is responsible for carrying the large majority of data used by smartphone users in most countries. If there is no Wi-Fi, mobile operators would have to invest large amount of capital in the networks to increase the data capacity or they would have to restrict their users' data usage. It is estimated that 150,000–450,000 new radio base stations would be required in order to cope with the traffic generated by world smartphones in the absence of Wi-Fi. That is why we can say mobile operators are saving yearly an investment of $30–$93 billion by utilizing the Wi-Fi technology and the license–exempt spectrum.

And if we look a bit further to 2016, the growth of data traffic (40% yearly growth) will require mobile operators to deploy an additional 115,000 base stations, which is about a 4% increase from today's numbers. In this case if we do not utilize the Wi-Fi network, the mobile operators would need to deploy additional 1.4 million base stations, which is about a 43% increase from today's numbers. The difference in cost between these two scenarios is around $250 billion, which is quite huge and may not be acceptable by the telecom industry. The only way they can fill the gap is to deploy more femtocells or picocells, but they are also much more expensive than the Wi-Fi deployments.

To facilitate the use of Wi-Fi networks by mobile data users, the cellular network operators have provided the mechanism to automate user connections to

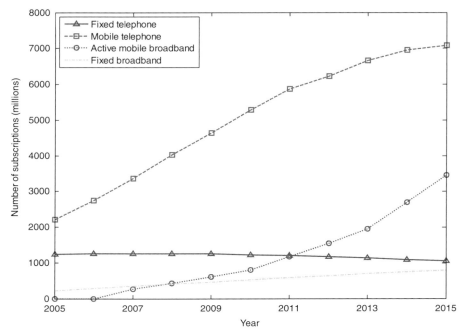

Figure 6.5 Subscription growth of various networks from 2005 to 2015. (*Source*: Data from https://www.itu.int/en/ITU-D/Statistics/Pages/stat/default.aspx.)

Wi-Fi networks. With these efforts by the telecom industry, we can expect there will be a further surge in the Wi-Fi network usage.

From Figures 6.5 and 6.6 we can see that the demand of using mobile data worldwide has been increasing with an incredible speed, especially in developing countries. Also, the place of using mobile data is moving gradually from outdoor to indoor in recent years, while the cellular network is originally designed for outdoor usage and voice service. It is obvious that cellular operators are relying on small cell deployment and offloading the traffic to the high-frequency Wi-Fi system to quickly adapt to this change. But for outdoor mobile users, what the operators doing now to satisfy users' demands is to either build more base station sites or increase the quantity of spectrum available. In fact, if the operator can push and promote other license–exempt systems that are also suitable for the outdoor environment to their customers as an alternative solution with additional low costs, it will be a win-win situation because the operator will be able to use more "free" spectrum and more cheap sites to serve their outdoor users.

Besides the widely used Wi-Fi technology, other license–exempt spectrum technologies also contributed to a lot of businesses and projects, which bring broadband access to many millions of people who are not in the coverage of fixed and mobile networks. Wireless backhaul is one of these successful businesses and projects. Because the cellular industry has been moving from building large-scale base stations to deploying many small-scale base stations, the backhaul network

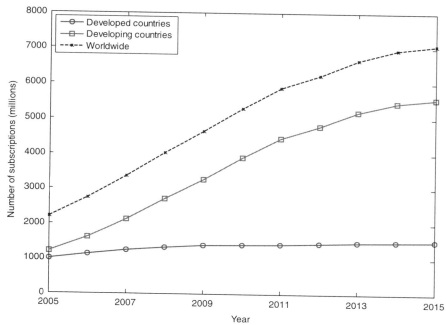

Figure 6.6 Comparison of mobile line subscriptions in different regions. (*Source*: Data from https://www.itu.int/en/ITU-D/Statistics/Pages/stat/default.aspx.)

connecting all the small cells becomes quite important. In addition to the traditional wired backhaul, using license–exempt spectrum as the backhaul becomes more and more common.

For wireless backhaul use case, the two main requirements are high-throughput and long propagation distance. The existing license–exempt spectrum technologies suitable for broadband delivery use higher frequency band to transmit data stream, and therefore are more easily blocked by obstructions such as buildings and foliage, which limits the backhaul use case. Fortunately, the TV white space spectrum uses the sub-gigahertz unlicensed bands that have excellent propagation ability and therefore can easily penetrate into obstructions. This unique characteristic could make TV white space successful on both the broadband delivery and wireless backhaul scenario. TV white space is also a precious gift for unconnected rural areas, because entrepreneurs or operators could use less-expensive, but reliable, radio equipment capable of operating on TV bands to deliver cost-effective broadband services in those areas.

In Europe, the economic potential and market value arising from a wise management of the digital dividend (cleared spectrum + TV white space), has attracted many studies in the past few years. The main studied use cases include DTT, broadcast mobile TV, commercial wireless broadband services, wireless broadband services for public protection and disaster relief (PPDR), services ancillary to broadcasting and program making, and cognitive technologies.

It has been illustrated that the UHF band used for terrestrial broadcasting is estimated to generate €750–850 billion in net present value (NPV) for the European economy [4]. Obviously, the excellent propagation and coverage characteristics of this band are the most attractive parts to the major stakeholders. It is commonly agreed in Europe that a part of the VHF spectrum (ranging from 72 to 92 MHz at a first stage) will be allocated to mobile operators, which is estimated to generate €63–232 billion NPV, measuring the benefits for a 20 year period.

Nowadays, not only human require communication, devices/machines also do. With the emergence of machine-type communication concept such as IoT and M2M, more and more applications that rely on sensors and control mechanisms are becoming available, which can make our life better. In fact, there are already more machine users than human users connected to the Internet at this moment. It has been reported that microcontroller (the heart of an IoT device) sales in 2014 are projected to grow 6%, to $16.1 billion, followed by increases of 7 and 9% in 2015 and 2016, respectively [5]. Shipments to distributors/retailers are also expected to post strong gains, climbing 12%, to 18.1 billion in the year of 2014. It is an unstoppable trend that the future devices must have both intelligence and connectivity parts embedded inside them.

There will be a lot of possibility and varieties of IoT applications, and some of them may bring incredible economic benefits. It is estimated that the number of intelligent connected devices is likely to exceed 100 billion by 2020 [1]. According to Metcalfe's law, the value of a network is proportional to the number of connections instead of the number of users. If this is true, then we can expect a huge economic boost on networks followed by the maturity of the IoT industry. It is estimated that the combined economic contribution (human and machine communication) by 2020 could reach $1.4–2.2 trillion per year, which is around five times that of Internet today.

Due to the mobility nature of the M2M device and the huge amount of devices required in IoT applications, the license–exempt spectrum is probably the most economical solution to carry out the data communication in the IoT application. It is estimated that there will be 1–2.5 billion machine connections using cellular system operating on licensed spectrum, while the others (97.5–99 billion) will be using license–exempt spectrum. This is because the license–exempt technologies are cost-effective, less-complicated, and easy to deploy and manage. Without the license–exempt spectrum and technologies, it is almost impossible to make Internet of things successful.

Among the license–exempt technologies, TVWS has huge potential to become the core technology to support the whole IoT businesses. The comparison of the license–exempt technologies that are often used in IoT applications is shown in Table 6.3. The propagation characteristics of sub-1 GHz spectrum make it possible for the TVWS technology to provide excellent coverage at low power consumption. No other license–exempt technologies like TVWS can fulfill simultaneously the two key requirements of IoT applications.

The importance of using the sub-1 GHz spectrum can be seen by comparing the smart electricity meter development in the United States and Europe. The

Table 6.3 Comparison of Different License–Exempt Technologies

	Wi-Fi	Bluetooth	ZigBee	TV White Space
Data rate	>5 Gbps	1 Mbps	250 Kbps	10 Mbps
Range	30–100 m	10 m	30–100 m	500–2000 m
Power consumption	Moderate	Very low	Low	Moderate
Network topology	Infrastructure/ Mesh	Infrastructure	Infrastructure/ Mesh	Infrastructure/ Mesh
Penetrating ability	Weak	Weak	Weak	Strong

United States allows the operators to deploy smart electricity meters using the spectrum at 900 MHz, while in Europe the operators are still not allowed to use that spectrum. As a result, the operators in Europe have to seek other solutions such as power-line communication or cellular systems, which did not bring satisfactory results to the operators.

Due to the freedom to use the spectrum and the easiness to build the network, more and more broadband applications using license–exempt spectrum are created with huge market potential. Wi-Fi broadband and its varieties are typical examples of creating successful markets. Also, there are many specialized networks being built using license–exempt spectrum to provide specific functions such as home/ building automation, factory monitoring, grid management, and smart city management. They are more resilient to failure than the traditional wide area network using fixed infrastructure or licensed spectrum as recovery from system failure is typically built into the networks.

Another important use case of the license–exempt spectrum is the rapid formation of the emergency communication network, especially when a natural disaster occurs. In such an emergency situation, the traditional wired infrastructure or cellular-based communication is often destroyed and hence not available, and it is always time-consuming to reestablish both types of networks. The license–exempt spectrum technologies, however, have the ability to being quickly deployed and hence can facilitate the disaster relief by resuming the communication in a short time. TVWS is especially suitable for use in disaster relief, not only because the TVWS signal can transmit over a long distance but also because the signal can penetrate into collapsed walls and obstacles. The value of TVWS for social aspects such as disaster relief cannot be purely measured by return-on-investment (ROI).

6.2 SPECTRUM TRADING AND MANAGEMENT

With the rapid growth of mobile data demand, it is often seen that operators need more spectrum to satisfy their users with specific requirements such as coverage, bandwidth, QoS, service duration, and price. Operators have several options while dealing with this issue. The first option is to optimize their currently owned spectrum for the efficient use of spectrum by having a better frequency planning, creating better coexistence strategies or reassign their own spectrum. However, this option is not always applicable since the spectrum could be already fully loaded and hence there is no room for such kind of rearrangement. Another possible option is to buy new spectrum with exclusive rights, for example, the new released TV spectrum, if the operator wants a nationwide deployment. However, only few operators will be able to apply this option because the cost for owning a spectrum is very high. Another moderate way to obtain additional spectrum is to temporarily own the exclusive right of the spectrum in the secondary market; TV spectrum is especially suitable in such kind of business.

The temporarily exclusive right approach is an attractive alternative and the market potential of this alternative could be very huge. It is obvious that this secondary spectrum market will largely increase the efficiency of allocating, assigning, and using the spectrum. The secondary spectrum market also offers opportunities for license holders to trade or lease their licensed spectrum for a period of time when the demand is much lower than the supply, or when the license holders no longer need the spectrum because of changes in technology, business strategy, or market share. For the case of TVWS, trades may occur while secondary users are not going to use the obtained spectrum in a specific area for a period of time. The trade in TVWS spectrum will reduce the complexity and overhead of the license-exchange procedure that always involves multiple transactions hosted by a central management entity.

Spectrum trading will benefit spectrum users of various types as mentioned below [6]:

- Large users of spectrum, such as telecommunications companies, will benefit from the potential to access more spectrum by trading for expanding technologies or services.
- Small users of spectrum, such as private business owner and end users, will benefit from the opportunity to generate profits from selling their spectrum that is temporarily unused, or from purchasing more spectrum at low cost to enhance their business.
- Business entity will have more opportunity to compete for spectrum for new technologies or services with the incumbents. Spectrum trading will also help to remove obstacles to enter into markets that were previously not friendly to new players because of the lack of access to spectrum.

Reference [7] shows that trading and modifying the terms of licensed spectrum is just stepping out as its first step toward having more market opportunities. Those initiatives must ensure that procedures are lean and simple, and there is sufficient information given to both the buyer and sellers. Efficient markets are those that generate the largest total welfare for both consumers and producers and where trades that increase social welfare should not be obsessed.

It is important to simplify the transaction procedure and lower the transaction cost to make the spectrum trading market successful, because the trade will only occur if an entrant values an existing portion of spectrum more than the incumbent user does and if the difference between the two is larger than the transaction cost. Buyers and sellers may need to learn to be confident about the real value of spectrum in order to really benefit from the spectrum trading. The most important information that can help both buyers and sellers to obtain a better understanding of the value of the spectrum they are trading is the actual price paid in similar transactions happening recently.

6.2.1 Primary Users' Incentives to Share the Spectrum

There is always a concern from the primary users who wants to sell their spectrum; the concern is whether or not the potential buyers will use their acquired spectrum to provide a service competing with the sellers or if the buyers will introduce distortion to the sellers [8]. Unless this concern is addressed, the primary users will be unwilling to sell or lease their spectrum.

There are also several basic steps to encourage primary users to share portions of their spectrum for secondary use [4]:

- Establish a framework for secondary markets
- Define and create the tradable usage rights and obligations
- Permit various forms of trading of these rights
- Establish rights to protection from interference and obligations not to create harmful interference in relation to liberalization of use
- Clarify rules on the expiry of usage rights and regulatory powers to reclaim them
- Develop clear rules to ensure effective enforcement of rights and obligations
- Establish and clarify enforcement mechanisms, whether regulatory or through private rights of action in courts.

6.2.2 Secondary Users' Incentives to Buy the Spectrum

It is always a concern from the secondary users that whether the spectrum they purchase or lease can really let them provide stable and satisfied service to their customers. So it is important for the regulation authority and the database operator to make it clear on the status of the band, including the available duration, the allowed transmission power, the coverage area, and also the expected QoS the band can provide.

There are multiple ways to carry out the spectrum trading in secondary markets. The trading may occur through intermediaries such as bandwidth brokers or through the process of online automated spectrum sharing and trading in a real-time fashion. The secondary users may need to indicate the types of services, access characteristics, or service levels to find a match on the primary user while searching for the available spectrum. The access type may need to further indicate whether this is a long-term lease, a scheduled lease, a short-term lease, or spot markets, because there are always different discovery mechanisms for different access types.

If the secondary spectrum trading becomes successful, it will help the secondary users to reduce their cost on providing services. It (secondary spectrum trading) is especially necessary when there are a large number of participants. The spectrum utilization for secondary use will be high if the data usage is biased toward continuous spectrum usage, which forces the market to drive secondary use transactions and suppress opportunistic behaviors of intermittent spectrum usage. Although spectrum users with both small and large coverage requirements can benefit from the expansion of license–exempt spectrum, the users with smaller coverage can indeed obtain greater benefits from an additional spectrum in the unlicensed band [8].

The spectrum trading and sharing described above are typically done via an automated system. This system performs the job of a regulator in a real-time and automated manner. We call these collectively as digital regulator.

6.2.3   Case Study: High Priority Channel in Singapore's TV White Space Regulation

IDA, the regulator in Singapore, has finalized the regulatory framework for TVWS operations in June 2014 [9] and shown in its previous study that most of the TV channels are underutilized in Singapore [10].

One major concern from the industries to IDA is the possibility of zero TVWS channel available for WSDs in time or geographic locations, as some applications may require more certainty in transmission. To address this concern, IDA proposed a new type of TVWS channel called the HPC, which is always available to WSDs. However, WSDB provider may impose fees to WSDs for using the HPCs. For a start, IDA has proposed to designate two HPCs for WSDs that are willing to pay for higher level of spectrum availability and access.

The fundamental task of the WSDB is to provide WSDs the information of available TVWS channels at their locations. With the introduction of HPCs, it creates a new business opportunity for WSDB provider to earn revenue from WSDs that do not mind paying fees for better QoS. In future, pay-per-use unlicensed channels may also become a common feature for spectrum sharing if the demand for better QoS in unlicensed channels becomes high. In the report by President's Council of Advisors on Science and Technology (PCAST) [11], it is recommended that secondary licensees can register and access WSDB for some QoS protections possibly in exchange for a fee.

The secondary user who purchases the HPC can also sell or lease the HPC to other secondary users if he/she does not require for a period of time. Undoubtedly, a matured secondary market for spectrum trading can stimulate the users to register for the use of HPC, and hence will help to bring more benefits to the WSDB provider.

6.3   POTENTIAL APPLICATION SCENARIOS

6.3.1   Wi-Fi with Cognitive Access to TV White Space

Conventional Wi-Fi operating in the unlicensed 2.4 GHz industrial, scientific, and medical (ISM) bands is short in transmission range, and yet it is a very successful technology in delivering cost-effective wireless Internet access to both homes, offices, and public areas [12]. Many researchers have investigated how to extend the transmission range of Wi-Fi so that it can obtain more users and reach its maximum revenue [13–15]. Different methods, including MIMO technology, high-gain antenna, and so on, have been proposed to stretch the range of current Wi-Fi system.

The first problem the 2.4 GHz ISM band faces while extending its transmission range is the NLOS transmission. In a long-range transmission scenario, it is very likely that the transmission path will be blocked by trees, vegetation, hills, or buildings. These obstructions will reflect and absorb the radio signals, and therefore the received signal will become very weak or in many cases even become totally lost. Another problem of 2.4 GHz ISM band is that it is very crowded, and thus will easily receive interference from other technologies and devices such as microwave ovens, baby monitors, wireless cameras, remote car starters, Bluetooth, ZigBee, residential wireless phones, and so on. There are several solutions proposed to resolve these two problems. For example, using towers to overcome obstacles and shifting to the less crowded 5 GHz band. Table 6.4 gives the summary of the current methods and solutions used to extend the range of conventional Wi-Fi networks. However, the propagation characteristics of the GHz bands still make it difficult for long-range transmission. Shifting to the sub-gigahertz bands is becoming inevitable for long-range Wi-Fi.

Compared to the 2.4 or 5 GHz ISM bands, the sub-gigahertz TVWS can transmit longer and is able to penetrate into obstruction easier due to its inherent propagation property. It also suffers less from interference as the TV bands are comparatively clean, which benefits a lot to the long-range and high speed wireless internet connection. In addition, Wi-Fi is a mature, well-understood technology that is inexpensive and easily available, Wi-Fi technology utilizing the sub-gigahertz will undoubtedly attract more attention from various application scenarios.

Perspective [12] estimates the economic value that might be generated from existing Wi-Fi applications switched to TVWS bands to be in the range of $3.9 –7.3 billion a year over the next 15 years in the U.S. market. Ofcom also believes

Table 6.4 Summary of Conventional Methods for Extending the Range of Wi-Fi [13–15]

Setback	Solution
Requirement for line-of-sight between endpoints	• Take advantage of terrain elevation • Avoid areas with obstacles • Use high towers to provide Fresnel clearance
Vulnerability to interference in the unlicensed band	• Operate in rural areas • Migrate to the less crowded 5 GHz band
Power budget limitation	• Use high-gain directional antennas
Timing limitation	• Modify the MAC mechanism
Cost and usability	• Use cheap antennas • Employ technology which is affordable and easier to use by people with limited training

that the net benefits to consumers could be worth £2–3 billion over 20 years for the United Kingdom. It can be foreseen that there will be more and more successful long-range Wi-Fi Internet services and these services will bring market potential in the following areas [4].

- Business
  - Provide coverage to a large office or business complex or campus
  - Establish point-to-point link between large skyscrapers or other office buildings
  - Bring Internet to remote construction sites or research labs
  - Provide cheap Internet in road, sea, and air transport systems

- Residential
  - Bring Internet to a home if regular cable/DSL cannot be hooked up at the location
  - Bring Internet to a vacation home or cottage on a remote mountain or on a lake
  - Bring Internet to a yacht or large seafaring vessel
  - Share a neighborhood Wi-Fi network

- Disaster areas
  - Bring quick, agile, and cheap Internet connections in areas devastated by disaster
  - Provide additional support in coordination of relief efforts

- Hospital applications
  - Access to patient records
  - A rapid voice over Wi-Fi communication system

In the United States Google, Microsoft, and Facebook have been leading several initiatives on the low frequency and long-range Wi-Fi technology, because they found it is highly attractive to adopt the technology for providing rural areas with high-speed Internet service. The substantial number of initiatives for adopting TV bands and using TVWS technology in the Wi-Fi industry is strong evidence for the viability of the market potential of long-range Wi-Fi over TVWS.

6.3.2   UMTS and LTE Extension over TV White Space

LTE is undoubtedly the largest and the most successful mobile broadband business in the world. What made LTE's success is the ability to enhance the cellular technology performance by utilizing the new spectrum and reusing the existing GSM spectrum. The mobile broadband operator can further benefit from the transition from analog to digital terrestrial television, which will release large amounts of spectrum potentially for mobile broadband use. There were already several countries such as Denmark, Finland, France, Germany, Sweden, Switzerland, and so on auctioning the 790–862 bands for the use of mobile broadband services (mostly LTE services).

Since LTE can operate on different bandwidth from 1.4 to 20 MHz, it gives operators very high flexibility in their commercial and technical strategies. While deploying at higher frequencies (always with higher bandwidth), LTE is capable of providing high-throughput networks; while deploying at lower frequencies, LTE is able to provide ubiquitous coverage for their user.

With the increasing demand for mobile data, operators are keen to have more bandwidths to increase their revenue to satisfy their users. Broadband subscriptions have been increasing dramatically these years, and two-thirds of these subscriptions will use mobile band. Based on this trend, several European countries have already taken some actions in these years to use the TV bands for mobile broadband services. The countries and their actions are listed as follows:

- Denmark, Finland, France, Germany, Sweden, and Switzerland have allocated 790–862 MHz for mobile broadband services.

- The UK government proposed ring-fencing of some digital dividend spectrum so that operators can push 3G beyond their license obligation of 80% population. In return, operators would receive an indefinite extension to their 3G licenses.

- In Germany, trials have been conducted by some local authorities in Mecklenburg-Vorpommern to use digital dividend frequencies to provide HSPA or LTE broadband services. The auction mentioned earlier of course helped to propel trials of this kind.

- In Italy, Communications Undersecretary also announced that frequencies vacated during the analogue switch off will be auctioned.

- Norway also announced that 790–862 MHz will be allocated to mobile broadband.

- In Spain, the Industry Ministry has issued a press release indicating digital dividend spectrum would likely be reserved for cellular use, especially for mobile Internet broadband, with the effect from January 2015.

It is also expected that LTE can benefit from numerous and affordable devices suitable for both developed and emerging markets and secondary markets in both urban and rural areas. For urban areas, because indoor radio signal penetration is difficult, especially in concrete buildings, the wireless broadband signal can hardly reach to home. Now allowing LTE operating on TV bands will resolve this difficulty because signals in TV bands can easily penetrate into the building.

In rural areas, because the population density is low so the base stations are deployed to provide coverage at low bit rates only. Using TV bands can reduce significantly the number of base stations while providing the same or even higher throughputs, based on the fact that the number of base stations required at 700 or 900 MHz compared 2 GHz is reduced by almost 65% for the same data rate and the same coverage [4]. In other words, operators can significantly reduce the capital expenditures (CAPEX) and operational expenditures (OPEX) while letting their users experience better services.

6.3.3 Digital Video Broadcasting for Handhelds (DVB-H) with Cognitive Access to TVWS

Broadcast mobile TV is a service that allows users to receive and watch multiple TV channels real time using their mobile devices. Multicast/broadcast mobile TV is much more efficient on sending out TV programs than unicast mobile TV is. This service requires dedicated mobile broadcast networks, which are currently available in only few countries such as Austria, Finland, Italy, and the Netherlands. The technologies involved in providing broadcast mobile TV service include DVB-T, DVB-H, digital multimedia broadcast (DMB), and MediaFLO. Europe is currently using DVB-H as its broadcast mobile TV technology.

The most suitable frequency range for mobile TV is 470–862 MHz as it provides a balance between range of coverage and antenna size. The frequency range also makes it possible for mobile devices to receive mobile broadband and mobile TV services at the same time. While operating at the mentioned frequency range, the broadcast tower must query the database first to obtain the permission to broadcast programs at certain channels, and may also need to adopt the cognitive radio technology to avoid interfering with the primary user. Moreover, the interactive mobile TV system that makes use of both mobile TV and mobile broadband systems has been attracting many interests. TVWS will have the chance to become the mobile broadband part technology, which is responsible for carrying the uplink data used in the interactive mobile TV service for its lower cost compared to cellular technologies.

From the business viewpoint, one of the biggest challenges to adopt DVB-H for mobile operators is the issue of business and revenue models. Mobile operators who wish to run the mobile TV service are expected to provide their original domain expertise such as service provisioning, billing, and customer care, and hence broadcasters or content providers would have ownership of the content and the overall visual experience. In this case, mobile operators would need to differentiate their offerings and provide value to ensure customer loyalty and remain profitable.

It is very likely in Europe that DVB-H services over mobile phones will trigger new television viewing behavior for consumers and create a new market for television viewership. However, spectrum access restriction ruled by national agencies could be a critical barrier to the adoption of DVB-H mobile TV services. As a result, TV white space could be utilized as the spectrum where DVB-H systems may operate.

Besides Europe, DVB-H is by far the most acceptable technology/standard worldwide in bringing the mobile TV service to market. Not only operators, but also broadcasters, broadcast network operators, and content providers are on the same page. The large number of consumer DVB-H trials currently taking place across the globe is a good indication. In Europe, trials have so far taken place in Germany, Finland, the United Kingdom, Spain, Switzerland, France, and the Netherlands. The outcomes of these trials are very successful and encouraging. For the trial in Finland, 58% of the users thought that mobile TV services would be popular and exactly half of the users thought that they will be willing to pay €10 per month for such services.

6.3.4 M2M Communications

The government-owned public applications such as environmental management (pollution/air quality monitors, weather stations, water level monitors), urban land-scape (street lighting control systems, parking meters), and health care (dialysis machines, defibrillators, ventilators, pacemakers) very much rely on the license–exempt spectrum because the devices carrying out these applications need to communicate with each other on these bands with minimum cost. Enhancing the link quality on these bands will significantly increase the efficiency and accuracy of these applications.

The number of machine-communication devices is now much more than the number of human-communication devices and will have an explosive growth in the near future. The license–exempt technologies such as Wi-Fi, Bluetooth, and ZigBee are quite suitable to support such a great number of connections with short trans-mission range. However, if the coverage requirement of the application is above several hundred meters or with obstacles in between, TVWS will become the most applicable technology to accomplish the mission.

We are going to estimate the economic value of M2M communications based on four applications: advanced meter infrastructure, security, energy demand side management devices, and telehealth.

Advanced meter infrastructure provides detailed, time-based information regarding the utilization of electric, gas, and water meters for both utilities and con-sumers. The gas/electricity/water meters transmit the collected data through a variety of communications technologies such as power line communication, fixed radio frequency, and public networks (landline or cellular) [16].

Security applications always need to establish the wireless link from entry points, appliances, lights, or Heating, Ventilation and Air Conditioning (HVAC) systems to a central hub of consumer's home. This is often accomplished using intermittent control signals spectrum, and the information carried includes control signals and occasional data signals. Such systems can be seen in homes, small busi-nesses, and large corporations. Wireless technology is preferred because it prevents the building from being damaged by installing cables inside the building. TVWS is even more preferred because its penetration ability can avoid blind angle of signal transmission.

The energy demand side management device such as the smart home power meter can reduce energy costs, helping homeowners to save money by using less energy. A two-way communication technology will allow smart meters to send real-time energy consumption information directly to the consumer in a house, motivat-ing them to conserve energy and save on their utility bills. The basic implementa-tion will allow the energy consumption information to be shown in consumers' PCs, tablets, or even smart phones so that consumers will be aware of the energy con-sumption in their houses.

The telehealth devices use the medical implant communication service (MICS) band radios with very low power consumption to communicate between implanted medical devices (cardioverter defibrillators, neurostimulators, hearing aids, and

Table 6.5   The US M2M Device Forecast (2013–2017) [3]

	2013	2014	2015	2016	2017
Total connections (in millions)	95.0	129.0	175.2	311.1	314.9
Connections rely on license–exempt spectrum (in millions)					
• AMI	7.1	9.8	11.9	18.2	27.8
• Security	7.8	9.9	12.7	15.2	18.2
• Demand side management	4.9	5.8	6.7	7.9	9.2
• Telehealth	2.7	4.4	6.9	8.5	10.6
Total connections rely on license–exempt spectrum (%)	47.8	48.7	48.2	50.7	54.1
Total M2M market (billion $)	32.6	36.9	41.7	47.1	53.0
Share of M2M market rely on license–exempt spectrum (billion $)	15.6	36.9	20.1	23.9	27.8

automated drug delivery systems) and nearby monitoring equipment. A simple model based on the forecast of M2M connections in the United States and the average traffic per device for the four applications are given in Table 6.5 [3].

6.3.5 Smart City Deployments and Applications

Smart city deployment involves distributed networks of intelligent sensors that can measure many parameters for more efficient management of the city. The data gathered from intelligent sensor is transformed into meaningful information, which is then delivered in real time to citizens or the appropriate authorities. There are many smart city-related applications enabled by wireless sensor networks. Telecom advisory services [3] gives some of examples as follows:

- Citizens can monitor the pollution concentration in each street of the city or they can get automatic alarms when the radiation level rises above a certain level.
- Municipal authorities can optimize the irrigation of parks or the lighting of the city.
- Water leaks can be easily detected or noise maps can be obtained,
- Vehicle traffic can be monitored in order to modify the city lights in a dynamic way.
- Traffic can be reduced with systems that detect the closest available parking space and inform motorists of the space's location. This information saves time and fuel, reducing traffic jams and pollution while improving the quality of life.

Smart city deployment could bring about economic impact such as job creation and industry development in areas such as sensor network deployment, data analysis, and value-added services. To further investigate the economic value that smart city deployment can bring, GOGEU breaks down the effects of smart city deployment could have [4]. These effects include the following:

- *Improved mobility:* Strong ICT infrastructure and sustainable transport systems
- *Economic growth:* High productivity, entrepreneurship and ability to transform
- *Sustainable environment:* Sustainable resource management, pollution prevention, and environmental protection
- *Quality of life:* Cultural facilities, housing quality, health and safety issues
- *Better administration:* Political strategies and perspectives, transparency and community participation in decision-making

Smart city deployment is targeting to improve the public transportation and quality of life in the city, and the improvement can further attract economic activity to cities and boost productivity. Making a city more attractive to live will motivate the labor to create more products and buyers to consume more of them. In addition, the market research firm IDTech estimates the global wireless sensor network market to be as much as $450 million, and that market could reach $2 billion in 2021. This results in a projected 2017 GDP contribution of $793 million in the United States [3].

6.3.6 Agricultural Automation

Modern-day agriculture requires a system-based approach for managing the crop production. The system-based approach, which relies on sensors for data gathering, can greatly enhance the efficiency of agriculture production. Different machines such as combine harvesters, field-spraying machines, fertilizer spreaders are equipped with different types of sensors to facilitate their functions. Since the machines used in agriculture production are always mobile, these sensors must use wireless technologies to send back their data to the control center.

Several technologies using license–exempt spectrum have been applied to the sensor network used in agricultural automation. Wireless meteorological stations and RFID are being used for the case of point-to-point communications, while the IEEE 802.15.4 with 6LoWPAN [17], Wi-Fi [18], and ZigBee are being used for the case of point-to-multipoint communications. TVWS is a potentially potent technology for agricultural automation because of its long transmission range and its ability to penetrate into plantations.

A study conducted by Robertson et al. [19] aimed at quantifying the economic contribution of precision agriculture. The study estimated that total benefits ranging from $13 to 28 per hectare can be achieved by improved fertilizer management and reduction of overlap of spraying. Also, Swinton and Lowenberg-DeBoer [20] estimated the net return of precision agricultural management at $47.01 per hectare.

In 2013, the United States had a total of 325 million crop acres where the precision agriculture has been adopted. Assuming the benefits of adopting agriculture automation are between $13 and $28 per hectare, the economic value of agricultural automation would range between $513 million and $1.1 billion.

6.3.7   Public Safety with Cognitive Access to TVWS

The frequency allocated for public safety in Europe is very limited, only the band 380–385 MHz and the band 390–395 MHz are available. These bands are therefore very congested. TVWS that uses dynamically allocated bands would be able to address this issue.

Communication is a crucial part of rescue operation when natural disasters occur. However, each entity in PPDR has their own radio-communication system, which makes it difficult to collaborate with each other. This problem was investigated by ITU-R and at the World Radio-Communication Conference in 2003 (WRC03) in its resolution 646, and an important agreement regarding to the harmonization of frequency band of PPDR was reached. They separated the world into three regions where corresponding bands can be used for PPDR as follows.

- *Region 1 (Europe and Africa):* 380–470 MHz with 380–385/390–395 MHz is a preferred core harmonized band
- *Region 2 (The Americas):* 746–806 MHz, 806–869 MHz, 4940–4990 MHz
- *Region 3 (Asia and Australasia):* 406.1–430 MHz, 440–470 MHz, 806–824/ 851–869 MHz, 4940–4990 MHz, 5850–5925 MHz

The existing PPDR services are mainly using either TETRA or TETRAPOL standard, which uses the band from 380 to 400 MHz. However, these networks only provide voice and narrowband services. Because users of the PPDR service are often on the move, having broadband access is an increasing demand for these users.

The core concepts in public safety communications are the following:[1] (1) *Operability:* The system must provide the requisite services across the spectrum of requirements for public safety. (2) *Interoperability:* The ability of disparate systems to operate with each other in a seamless fashion. (3) Reliability: Activities including error prevention, fault detection and removal, and measurements to maximize reliability. (4) *Resiliency:* The ability of a communications system to respond to change and/or to recover from mishap is crucial in establishing the resilience of the system. (5) *Redundancy:* The functionality of a communication system that is achieved by alternative means. (6) *Scalability:* Ability to adapt and grow with expanded requirements and users. (7) *Security:* End-to-end integrity of the transmission process to insure error free and uncompromised exchange of information. (8) *Efficiency:* Optimize the resource use.

TVWS that has excellent propagation and dynamic spectrum access characteristics will be able to address the above concerns. With TVWS technology, white space database operators can easily implement a prioritizing mechanism that assigns extra spectrum for a specific disaster area based on spectrum availability information provided by the white space database.

---

[1] https://www.fcc.gov/help/public-safety-tech-topic-7-core-concepts-and-software-defined-radio-public-safety.

6.3.8 PMSE with Cognitive Access to TVWS

Bands of 470–790 MHz have already been used for PMSE applications for many years. The usage schemes on these bands are different from countries to countries. After the digital switch over and of the digital dividend, the available bands for PMSE applications were reduced. However, it is claimed within CEPT [21] that PMSE still requires more bandwidth in case that a major event occurs when no general spectrum is capable of handling the event.

While being used for PMSE applications, spectrum needs to be managed and planed very carefully to minimize interference. The major event could be either a planned event or unplanned event. In case of an unplanned event, timely coordinating the frequency usage to avoid interference is a big challenge. Even if it is a planned event, finding exclusive spectrum is also very difficult because the usage of spectrum always coexist with other usages for other services and applications.

Although PMSE services have already used white spaces bands for years, the cognitive access part has still not been implemented in an automatic fashion. With human-involved cognitive control, the spectrum margin used to protect the service from being interfered will be very large due to the lack of information. With the automatic mechanism that gathers the additional information automatically and periodically, a smaller spectrum margin that helps to increase the spectrum efficiency can be obtained.

Currently, the PMSE devices are still using analogue transmission because of the lack of digital equipment. Therefore we cannot expect the device will soon have the cognitive access module as a fully integrated product. This fact must be taken into account while calculating the economy that PMSE services could bring.

It is estimated that the number of users is around 4–5 million across Europe according to the PMSE industry [21]. According to the German industry, the number of Professional Wireless Microphone Systems (PWMS) is estimated at 600 to 700 K. Microphones in Germany generally operate in the band 790–862 MHz. These are all the potential market for TV white space in PMSE services, especially when the cognitive access technology will be really required someday.

6.4 SUMMARY

As the reader could see, the potential application of TVWS is huge and is perhaps only limited by our imagination. This is echoed by the then FCC Chairman Kevin Martin: "Opening the white spaces . . . will have access to devices and services that they may have only *dreamed* about before."

REFERENCES

1. R. Thanki, *The Economic Significance of Licence-Exempt Spectrum to the Future of the Internet*, Wireless Innovation Alliance (WIA), 2012.
2. R. Thanki, *Measuring the Local Impact of TVWS Broadband*, PN ECS Partners, 2015.
3. R. Katz, *Assessment of the Future Economic Value of Unlicensed Spectrum in the United States*, PN Telecom Advisory Service, 2014.

4. COGEU, Cognitive Radio Systems for Efficient Sharing of TV White Spaces in European Context – D2.1: European TV White Spaces Analysis and COGEU Use-Cases, FP7 ICT-2009.1.1, Contract No. INFSO-ICT-248560, March 2010.

5. I. Scouras, Microcontroller Market Resurges, EE Times, August 2014. Available at http://www.eetimes.com/document.asp?doc_id=1323578.

6. OECD, The spectrum dividend: spectrum management issues, DSTI/ICCP/TISP (2006) 2/FINAL, November 8, 2006. Available at http://www.oecd.org/dataoecd/46/42/37669293.pdf.

7. P. Crocionia, Is allowing trading enough? Making secondary markets in spectrum work. *Telecommun. Policy*, 33 (8), 451–468 (2009).

8. A. Tonmukayakul, An agent-based model for secondary use of radio spectrum, Ph.D. thesis, University of Pittsburgh, 2007.

9. IDA, Regulatory framework for TV white space operations in the VHF/UHF bands, Decision Paper, June 2014.

10. M. H. Islam, C. L. Koh, S. W. Oh, X. Qing, Y. Y. Lai, C. Wang, Y.-C. Liang, B. E. Toh, F. Chin, G. L. Tan, and W. Toh, Spectrum survey in Singapore: occupancy measurements and analyses, in Proceedings of the IEEE International Conference on Cognitive Radio Oriented Wireless Networks and Communications (CROWNCOM), May 2008, pp. 1–7.

11. Executive Office of PCAST, Realizing the Full Potential of Government-Held Spectrum to Spur Economic Growth, Technical Report, July, 2012.

12. R. Thanki, The economic value generated by current and future allocations of unlicensed spectrum, Final report, Perspective Associates, September 8, 2009.

13. B. Raman, and K. Chebrolu, Experiences in using Wi-Fi for Rural Internet in India, IEEE Commun. Mag., Special Issue on New Direction in Networking Technologies in Emerging Economies, January 2007.

14. K. Chebrolu, B. Raman, and S. Sen, Long-distance 802.11b links: performance measurements and experience, MOBICOM, September 2006.

15. R. Patra, S. Nedevschi, S. Surana, A. Sheth, L. Subramanian, and E. Brewer, WiLDNet: design and implementation of high performance Wi-Fi based long distance networks, USENIX NSDI April 2007.

16. P. Adke, J. Bumanlag, B. Edelman, and U. Doetsch, *Spectrum Needs for Wireless Smart Meter Communications*, University of Colorado, Boulder, 2011.

17. M. Alberts, U. Grinbergs, D. Kreismane, A. Kalejs, A. Dzerve, V. Jekabsons, N. Veselis, V. Zotovs, L. Brikmane, and B. Tikuma, New wireless sensor network technology for precision agriculture, International Conference on Applied Information and Communication Technologies, April 2013, pp. 25–26.

18. M. Keshtagary, and A. Deljoo, An efficient wireless sensor network for precision agriculture. *Can. J. Multimed. Wirel. Netw.*, 3 (1) (2012), pp. 1–5.

19. M. Robertson, P. Carberry, and L. Brennan, *The Economic Benefits of Precision Agriculture: Case Studies from Australian Grain Farms*, Grains Research and Development Corporation, 2007.

20. S.M. Swinton and Lowenberg-DeBoer, Evaluating the profitability of site specific farming. *J. Prod. Agric.*, 11, 439–446 (1998).

21. CEPT REPORT 32, Recommendation on the Best Approach to Ensure the Continuation of Existing PMSE Services Operating in the UHF (470–862 MHz), Including the Assessment of the Advantage of an EU-level Approach, October, 2009.

Chapter 7

Future Development

TVWS has attracted a lot of interest in the past 5 years. While there were initial successes, the scales were limited. This could be due to the nascency of TVWS ecosystem. In particular, regulatory uncertainty and absence of clear business model are holding back large-scale deployments. In this chapter, we look at future developments in regulations and businesses that could potentially change the landscape. We also discuss technological trends in spectrum sharing and additional applications that could leverage on TVWS and other spectrum sharing technologies.

7.1 REGULATION

As discussed in Chapter 1, TVWS is one way of spectrum sharing technology. Besides TVWS, there are multiple alternatives where spectrum could be utilized in a shared basis. The Wireless Innovation Forum defined multiple tiers of spectrum sharing as indicated in Table 7.1 [1].

- Level 0: *Exclusive Use*—Spectrum is assigned on an exclusive basis to a primary holder of spectrum rights (primary user) across the regulatory region.

    This establishes an incumbent (or primary user (PU)) and can include, for example, cellular operators and radar installations. The expectation is that there is no interference.

- Level 1: *Static Spectrum Sharing*—Exclusive use spectrum is shared by primary users on a geographic basis and not on a temporal basis.

- Level 2: *Managed Shared Access*:
    - Level 2A: *Industry Managed*—Unused exclusive use spectrum in a specific location may be leased by the primary user to a third party on a temporary basis (SU). SUs at this level are protected, exclusive users for the assigned period.

        Policies/rules under which such arrangements can occur are set through negotiations between the PU and SUs following regulatory requirements established for such activities, and such rules may require the SU to clear the spectrum under specific conditions should the primary user require the spectrum.

TV White Space: The First Step Towards Better Utilization of Frequency Spectrum, First Edition.
Ser Wah Oh, Yugang Ma, Ming-Hung Tao, and Edward Peh.

Table 7.1   Tiers of Spectrum Sharing

Level 0: Exclusive Use

Level 1: Static Spectrum Sharing

Level 2: Managed Shared Access

Level 2A: Industry Managed

Level 2B: Government Managed

Level 3: Dynamic Spectrum Sharing

Level 3A: No Priority Access

Level 3B: Priority Access

Level 4: Pure Spectrum Sharing

Level 4A: Lightly Licensed

Level 4B: Unlicensed

Examples: Licensed-shared access (LSA) and lease management agreements (LMA).

- Level 2B: *Government Managed*—Exclusive use spectrum in a specific location may be assigned by a regulatory agency on a temporary basis to a third party (SU). SUs at this level are protected, exclusive users for the assigned period. Primary users who are using the spectrum may be required to vacate for the assigned period.

Policies/rules under which the SU operates are set by the regulatory agency.

- Level 3: *Dynamic Spectrum Sharing.*
- Level 3A: *No Priority Access*—Spectrum access is nonexclusive. Spectrum held by a primary user that is not being utilized in a specific location and at a specific time is available for use by SUs on a first-come first-served basis so long as they do not interfere with the primary user. Such secondary use is unprotected, and the SU must vacate the spectrum when required by the primary user. There is a management function, via a database or other means, that ensures noninterference with the primary user and such management functions may be used to support coexistence between SUs.

Example: TVWS, 5 GHz U-NII.

- Level 3B: *Priority Access (Three-Tier Model)*—Spectrum access is nonexclusive. Spectrum held by a primary user that is not being utilized in a specific location and at a specific time is available for use by a SU so long as they do not interfere with the primary user. Certain SUs are assigned priority access privileges. Prioritization can be made based on multiple models (cost-based/micro auctions, public good, social factors/uses, FIFO, etc.). Access by priority users is protected, while access by all other SUs is not protected; priority users have first rights to available spectrum, and other SUs must vacate if a priority user wishes access. There is a management function, via a database or other means, that ensures noninterference with the primary user, manages access by priority users, and such management functions may be used to support coexistence between SUs.

    Example: 3.5 GHz Citizens Broadband Radio Service (CBRS), Singapore TVWS.

- Level 4: *Pure Spectrum Sharing.*
- Level 4A: *Lightly Licensed*—Spectrum is not assigned to a specific primary user. Use of spectrum is protected while occupied. Rules may exist for length of time spectrum may be occupied.
  Example: U.S. 3650–3700 MHz band.
- Level 4B: *Unlicensed*—Spectrum is not assigned to a specific primary user. Use of the spectrum is completely unprotected. Spectrum is available to any network or user within limitations/rules/policies established for each band.
  Example: ISM bands.

From the different tiers described, one could see that while TVWS is generally classified as Level 3A, certain hybrid combination is possible, such as the TVWS regulation in Singapore that adopted a combination of Level 3A and Level 3B concepts. In a more general setting, we also should not rule out other combinations of spectrum sharing.

7.1.1　Citizens Broadband Radio Service (3.5 GHz)

The overwhelming popularity of mobile devices continues to push the capacity of the radio spectrum. As a result, spectrum sharing is being considered not only for TV bands. The experiences and concepts of utilizing TVWS can be extended to other white space bands.

In 2012, the FCC released a preliminary proposal for a Citizens Broadband Radio Service (CBRS) that is based on a spectrum-sharing scheme using up to 150 MHz of spectrum in the 3.5 GHz band. Instead of being sliced into "static" blocks and sold to the highest bidder, the spectrum would be shared dynamically according to needs or demands. Under the scheme, the spectrum will be divided into three tiers allocated to either federal or nonfederal incumbents (PUs); priority access licensees (PALs); and an unlicensed tier for general authorized access (GAA). This concept is similar to the Singapore TVWS regulation where PALs is like HPCs. However, the detailed approach may vary depending on implementation.

Besides 3.5 GHz, there are a number of unlicensed frequency fragments in 5, 60 GHz, and so on. There are great potentials to utilize cognitive radio technology in these white spaces too.

7.1.2　Spectrum Refarming and Trading

The purpose of spectrum management is to give access to spectrum for the largest possible group of interested parties in due time, while ensuring the overall efficiency of spectrum use and avoiding harmful interference between the users.

In 2002, the Electronic Communications Committee (ECC) of European Conference of Postal and Telecommunications Administrations (CEPT) released a report "Refarming and Secondary Trading in a Changing Radiocommunications World" [2]. Although this report is issued by CEPT, the implementation of refarming and spectrum trading processes remains a strictly national issue.

Refarming in the traditional sense means the recovery of spectrum from its existing users for the purpose of reassignment, either for new uses, or for the introduction of new spectrally efficient technologies. As such refarming is a spectrum management tool that can be used to satisfy new market demands and increase spectrum efficiency. Refarming could be done voluntarily or by force. With refarming, the overall utilization and capacity is expected to increase.

Spectrum trading, on the other hand, is complementary to market-based spectrum pricing in the form of auctions or administrative incentive pricing and also to spectrum planning and regulation. It offers advantages of dynamic optimization of spectrum distribution including spectrum refarming. Taking down GSM network and free up the spectrum for 4G services is a typical example of spectrum trading. Development of trading will be promoted by certainty about licensees' rights, including freedom from interference, security of tenure and expectation of renewal, and flexibility to change the use of spectrum within the constraints of spectrum planning and international harmonization.

Spectrum trading is not unique to Europe. Recently, the Telecom Regulatory Authority of India (TRAI) also proposed spectrum trading mechanism, which is similar to the licensed shared access (LSA) for all bands.

7.1.3 Sharing in Licensed Bands

Besides licensed exempt bands, there are also opportunities for sharing in the licensed bands. Imagine if a regulator has nine channels to allocate to three operators. Should the regulator allocate three channels each or should the regulator allocate two each and keep three as common pool? The key idea is to allocate spectrum sufficient for operators to serve say 2- or 3-sigma (95 or 99.7%) of traffic. This will ensure operators have sufficient bandwidth to serve their customers majority of time. Whenever there is a surge in demand, the operator will dynamically request for additional licensed bands from the regulator or a neutral party appointed by the regulator in real time.

On a related trend, operators are starting to share their site infrastructure be it tower or even common antenna system (CAS) in order to save cost and speedup deployment. Moving forward, there will be more and more small-cell deployment where site acquisition and setup becomes even more tedious and costly. By collaborating in site deployments, for example, operators sharing the same site and same head-ends, they differentiate in the centralized service such as Cloud-radio access network (C-RAN). With this scenario, it makes more sense to allow common frequencies so as to maximize spectrum utilization. This concept is a variant of Level 1 sharing and will need regulators and operators coming together to realize it.

One of the mostly asked questions about TVWS or any other spectrum sharing technology is "will the spectrum used up when I need it the most?" This is indeed a key concern since most spectrum sharing policies only protect the PUs, such as broadcasters, but not the SUs that use the shared spectrum. Regulators have dilemma of protecting PUs while maximizing utilization of spectrum through secondary shared access. A good compromise is to have some levels of spectrum certainty for shared access while not affecting the PUs. For example, 2–3 frequency bands are identified in each geographical regions and time to guarantee spectrum availability for shared access. In fact, some forward-looking regulators such as Singapore IDA have already started this provision in their regulation. In the Singapore TVWS regulation, two channels are reserved as high-priority channels (HPCs) (see Section 4.4.1.3). Although the detailed access mechanisms of HPCs are not yet available, it guarantees minimal spectrum availability. Recently, FCC also released a notice of proposed rulemaking (NPRM) on "amendment of Parts 15, 73, and 74 of the commission's rule to provide for the preservation of one vacant channel in the UHF television band for use by white space devices and wireless microphones." In this NPRM, the FCC tentatively concluded that it will preserve a vacant channel in each area for use by WSDs and wireless microphones. This move will give TVWS a boost as more applications become possible with spectrum certainty.

7.1.4 Spectrum Sharing for IoT

Another key area of interest for TVWS or other spectrum sharing technologies is Internet of Things (IoT). While the large number of bandwidths allocated for TVWS is a good news from capacity perspective, there is a cost issue that one needs to consider. The 300–400 MHz frequency allocated for TVWS imposes a big challenge for a low cost design while trying to cover the full band. For typical applications to serve rural Internet, this additional cost is not a major issue. However, for IoT devices where the target cost is in the order or $30 or lower, support of full 300–400 MHz bands will unnecessarily increase the cost of the device, although the IoT devices do not require such a large bandwidth. One possible solution is to make use of spectrum certainty described above or its variants to lower the cost of IoT devices. To enable this, some universal spectrums are allocated for IoT accesses. With this in place, the IoT devices just need to cover a small bandwidth that will greatly reduce the cost of the devices.

7.2 TECHNOLOGIES

When cognitive radio (CR) was first conceptualized, the radio was intelligent enough to determine the radio environment, learn, and act accordingly.

While the first CR technology, TVWS, has achieved some early successes, the method of obtaining vacant spectrum for communication deviates significantly from its original idea of using spectrum sensing. This is partly because spectrum sensing technology could not achieve the stringent requirements yet. As a compromise, both technically and politically, white space database (WSDB) is adopted. With WSDB, the deployment becomes less flexible since the WSD (at least one device in a network) needs to have access to the Internet in order to query the available channels. In addition, for certain networks where connection to the Internet is prohibited, either due to security or other concerns, this architecture may not work.

There are many studies that compare WSDB and spectrum sensing. In reality, WSDB is more a policy tool rather than a new technology, while spectrum sensing is a technological solution. When technology fails to deliver the desired specifications, policies may come in to regulate. When technology such as spectrum sensing could achieve the desired performance in the future, policies can be relaxed. This is similar to economy where the government will let the market drive the behavior since the market is often more efficient. Regulation comes in only when market fails.

In this section, we discuss some possible technological trends.

7.2.1   Spectrum Sensing

Although spectrum sensing is not mandated by any regulation (U.S. FCC has spectrum sensing as an option), it does play some roles as below:

- Co-existence between SUs
- Additional data for improving WSDB
- When policy (WSDB) fails

7.2.1.1   Coexistence Between Secondary Users

WSDB only protects the PUs. It does not "guarantee" the "available channels" returned by the WSDB are clean for WSDs to use. Although there are some proponents who are looking forward toward collecting access information and expanding the role of WSDB to give some indication of the quality of the available channels, this proves to be difficult due to nonregular and random access patterns of the WSDs. A more straightforward solution is to have the WSDs sense the spectrum in a regular basis and determine which channels are more suitable for their needs, similar to "listen-before-talk" politeness policy adopted in modern Wi-Fi technologies. In short, WSDB protects the PUs and let WSDs know which channels are not occupied by the PUs, while spectrum sensing ensures coexistence between the SUs.

7.2.1.2   Additional Data for Improving WSDB

The current regulation for WSDB is conservative. It creates large protection zones and does not differentiate between outdoor and indoor accesses. It is also an estimate since most regulations use propagation model to predict the availability of

channels (the UK Ofcom has a slightly different approach where they provide channel availability based on broadcast data collected all over the years). This inevitably
results in inefficiency of spectrum utilization. The propagation model also could not
keep up with the constant change in geographical landscape.

Spectrum sensing, on the other hand, provides real time information about the
spectrum condition in a particular location. If this information is sent to the WSDB,
it could complement the propagation models used and the WSDB provider could
fine tune the estimation of available channels. For example, when Longley–Rice
propagation model is used, some parameters such as terrain profile require setting
of values based on data collected, through experiments or inference. These "values"
are either inaccurate in some settings due to lack of data or change over time due to
change in man-made terrain profile.

With real-time spectrum sensing and reporting to WSDB, the WSDB will be
able to calculate a more accurate prediction of PU propagation and thus improves
its protection to the PUs and at the same time improves overall spectrum utilization.
To achieve this, some regulation enhancements are required as well.

7.2.1.3 *When Policy Fails*

WSDB is currently the *de facto* method used for TVWS. There is a little argument
in using such a method to protect the PUs and let the SUs access available spectrum
in an opportunistic manner. This policy should work in developed countries where
regulations are clear and the enforcement is strong. The same may not be true in
developing countries where the regulator may have limited tools and resources to
enforce such regulation. Therefore, spectrum sensing becomes important in such
scenario where technology is used to avoid misuse of TVWS that could result in
interference to the PUs. For these countries, the results obtained from WSDB may
be verified automatically through a spectrum sensing module running in the WSDs.
Only when both indicate availability, then only the WSDs are allowed to use the
vacant spectrum.

The authors believe that DSA such as TVWS will follow similar path as typical economic principle. Due to uncertainty and the need to ensure all stakeholders
are comfortable, policy approach such as WSDB is required. When people started
to accept such dynamic use of spectrum, technology such as spectrum sensing will
come in to aid WSDB so as to ensure a better efficiency in terms of spectrum utilization. When the market (technology) is mature enough, the market (technology)
will take over.

7.2.2 WSDB

The current WSDB is still in its early stage. It is an experiment on a new way of
spectrum allocation in a dynamic fashion. However, the concept is promising as it
allows the regulators to adjust regulations quickly without having to recall and
reprogram WSDs that are deployed widely. In this section, we discuss the various

improvements possible with WSDB and lay some foundations for future evolution of WSDB.

7.2.2.1 Location-Adaptive WSDB

Typically, PU and SU cannot access the same spectrum simultaneously in a PU protection contour calculated according to the weakest PU received signal. Even for adjacent channels, the SU is only allowed to transmit at a lower level.

The current WSDB returns a list of available channels no matter whether the WSD is near the broadcast tower or far from it. By making use of this additional information, there is an opportunity to further improve utilization of spectrum.

In typical wireless communication, it is the signal-to-noise ratio (SNR) that determines the quality of the link. For TVs located near broadcast tower, the received signals are very high. Therefore, they can tolerate much higher noise levels. By making use of this property, WSDs could theoretically emit higher power at adjacent channels near this region without affecting the reception quality of the TV signals. For example, for WSD meeting say class 4 (Table 2.24), it is allowed to transmit at the lower allowed power when it is near the edge of the coverage of the broadcast tower, while the power could be increased to say class 3 or class 2 levels when it is near the center of the broadcast tower. Since WSDB knows the locations of the broadcast towers as well as the locations of the WSDs, there is potential for WSDB to coordinate and further enhance spectrum utilization in order to maximize total capacity.

7.2.2.2 Future Evolution of WSDB

While WSDB is currently used for TVWS and other DSA systems, it could be easily extended to include licensed bands and other types of spectrum sharing (refer to Table 7.1) as well. It could be viewed as "digital regulator" where this *digital regulator* allocates spectrum on a real-time basis automatically following some predefined rules instead of the current approach of manual allocation. The advantages of *digital regulator* are as follows:

- Spectrum could be allocated in a short period of time instead of one time allocation of 10 or 20 years, thus allowing more flexibility in spectrum usage.
- Spectrum could be allocated to two or more operators at the same time provided their geographical coverages are different, thus improving spectrum utilization.
- Spectrum auctions can be carried out in real time and more frequently to reflect the market value of the spectrum.
- Regulation and exceptions could be updated on-the-fly easier especially for temporary change in allocation due to special events.

With real-time allocation, spectrum utilization will improve and therefore spectrum crunch issue could be alleviated.

7.2.3 Antenna

In Chapter 4, we mentioned that antenna is an important component for efficient utilization of TVWS. This is because TVWS covers both VHF and UHF, which ranges from 54 to 806 MHz. The frequency range is too large for conventional antenna to achieve good efficiency with practical size.

Ultra-wideband antenna is often being considered. However, ultra-wideband means lower Q factor, and low Q factor means low out-of-band interference rejection rate and high return loss, which in turn sacrifice the performance of TVWS system.

In fact, a WSD's signal bandwidth is much smaller than its possible frequency range. Thus, tunable antenna could be a better solution for TVWS. With tunable antenna, the antenna impedance bandwidth is just the signal bandwidth, which guarantees good antenna Q-factor while the central frequency of the antenna can be tuned in full TVWS frequency range.

7.2.4 Related Technologies

Besides the common technologies being investigated for DSA and TVWS, there are other related technologies that may impact how DSA or TVWS works. We discuss some of these technologies in this section.

7.2.4.1 Cognitive Direct Conversion

One of the main applications of TVWS is for rural connectivity. In a typical setup, TVWS is used as the "middle-mile" to bridge Wi-Fi to the Internet as depicted in Figure 7.1. The interface between TVWS station (STA) and Wi-Fi router is via RJ45 Ethernet. In this configuration, signal received by the TVWS STA goes through RF conversion, analog-to-digital conversion, layer-1 baseband processing, medium access and networking layer before the data are being sent to the Wi-Fi router. In the Wi-Fi router, the reverse process happens before Wi-Fi signal is sent out from the Wi-Fi router. The exact opposite happens for the uplink communication. These many levels of processing and conversions result in additional latencies, overheads, and unnecessary processing.

Figure 7.1 Typical setup for rural Internet.

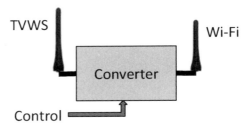

Figure 7.2   Cognitive direct converter.

A better approach is to combine TVWS STA with Wi-Fi router into one called cognitive direct converter as depicted in Figure 7.2. With this architecture, the TVWS signals received go through RF conversion directly to Wi-Fi signals without going through digital processing, and the reverse happens in the uplink. The control signal communicates with the TVWS AP and provides information on which frequency the TVWS channel is using. With this architecture, the latency is much lower and there is no additional overhead. Due to its simplified architecture, the cost of this converter is also much lower.

7.2.4.2 Full Duplex

In a two-way communication, typical systems adopt either frequency division duplexing (FDD) or time division duplexing (TDD). The former (e.g., 3G W-CDMA) differentiates the uplink and downlink transmission in different frequency bands while the latter (e.g., 2G GSM and Wi-Fi) achieves the same using time.

Researchers at the Stanford University broke this barrier with in-band full-duplex technology [3]. It allows transmission and reception of RF signals at the same time using the same frequency band. While the potential capacity gain of full duplex in a typical link is only twice, not as attractive as other innovative solutions such as MIMO and OAM, it is significant when it comes to DSA and TVWS.

For TVWS, the system has to constantly sense the spectrum to understand its utilization so as to avoid interfering with other systems. For TVWS operating in TDD mode, although sensing is possible during transmission blanks, the system has to be designed with additional quiet period for such sensing. Adding quiet period will reduce the throughput of the system. With full duplex, sensing can be done even when the system is transmitting. Thus, maintaining the throughput of the system.

In contrast for TVWS operating in FDD mode (not favorable since the uplink and downlink frequency pair cannot guarantee the same separation distance and thus more complex transceiver design is required), the system requires a sensing circuit operating in the same frequency as the transmitter and the sensing could only be carried out when the transmitter is not transmitting. This scenario is rare for the downlink of an FDD system as a typical FDD system is designed with downlink continuously transmitting control information. With full duplex, this restriction can be avoided since the system could continuously transmit signals while sensing is being performed at the same time.

Another application of full duplex is in mesh networks. Typical mesh networks suffer from drop in throughput for every hop they extend. With full duplex, there is no reduction in throughput since the transmission and reception can be performed at the same time. This advantage is applicable for all mesh networks, not only DSA or TVWS. However, the advantage is more prominent for DSA and TVWS as it allows DSA and TVWS to operate in mesh mode while constantly monitoring the condition of the spectrum.

7.2.4.3 RF OAM

Like all of other wireless frequency bands, TVWS will also face the increasing number of users. The band will become crowded when WSDs are widely utilized. How to further increase the spectrum utilization efficiency is still an issue.

Orbital angular momentum (OAM) for communication capacity boosting is a new area. OAM is a new kind of electromagnetic wave (EMW) propagation for communications where spin angular momentum (SAM) was used since wireless communications were invented in the nineteenth century. OAM is also called twisted wave. The twisted wave includes infinite modes that can coexist in the same space. Since different modes of OAM wave have different fingerprints over air, it is possible for one to modulate different information over different OAM modes but transmit them at the same carrier frequency in the same time and same space. Thus, OAM could increase the wireless spectrum utilization greatly. Increasing link capacity by OAM has been achieved in optical free-space and fiber communications. Now researchers are making effort to extend multimode RF OAM to wireless communication, where the spectrum is crowded and increasing spectrum utilization is extremely important. Unlike MIMO, which relies on antenna array and digital signal processing to enhance the system throughput, RF OAM aims to increase the system capacity in RF level without heavy signal processing. This is important for backhaul or fronthaul applications, where the data rate is ultra-high, and complex signal processing may not be fast enough to meet the speed requirement.

The targets of RF OAM for enhancing the performance of cognitive radio and TVWS is therefore centered around its capability to dynamically program the number of OAM modes and control its direction according to knowledge the system learns from the environments.

7.2.4.4 Software-defined Radio

Currently, operators share some of the infrastructures wherever possible, for example, tower sharing, common antenna system, and so on in order to save cost. Moving forward, with the trend of moving towards small cells, operators might increase sharing in order to reduce deployment cost. It is foreseen that operators may share the antenna or RF frontend, digitize as early as possible and send the digital signals to cloud radio access network (C-RAN) for processing.

To enable this, software-defined radio (SDR) is required. SDR digitizes analog signals either at carrier frequencies (less common due to high sampling rates required) or low intermediate frequencies (low IF). Advantages of doing so include flexibility in changing the operating frequencies, sharper filters in digital domain, and potential future cost down due to Moore's law.

With SDR, an operator has the flexibility of configuring the system to operate using licensed exempt bands, for example, TVWS, whenever possible. Whenever the license–exempt bands become unavailable or congested, the operator could reconfigure the system to use licensed bands. The operators also have the flexibility of using one or both licensed and license–exempt bands individually or concurrently.

7.2.4.5 WATCH

New technologies detecting PU receiver status can make a SU work in PU protection contours at the same frequency of PUs. Researchers from Rice University have developed a system that allows wireless data transmissions over TV channels while the TV is on [4]. This technology is called Wi-Fi in Active TV Channels, or WATCH, and received FCC approval to test it. The basic principle of WATCH is to actively monitor whether a nearby TV is tuned to a channel to avoid interfering with its reception. Some special techniques of signal cancellation are also used to insert wireless data transmissions into the same channel, which eliminates TV broadcasts from interfering with the WSD.

7.2.5 Privacy and Enforcement

The introduction of secondary access to coexist with the PUs could potentially open up a threat to the PUs as there will be more hardware readily available for hackers to exploit. Even from the software WSDB perspective, by mining the data, there is potential for the hackers to know the parameters of the PUs that could include military users. These are important topics to look into in order to ensure confidence in secondary access.

7.2.5.1 Privacy

WSDB stores the information of the PUs, often including the location of the transmitter, power, antenna heights, and so on. If a hacker performs multiple queries, there is a chance that he may gather enough information to derive the operating parameters of the PUs. This may be a potential threat to the PUs, for example, jamming of PU network. To increase privacy of PUs, the WSDB may need to secure, randomize the data, or monitor query patterns of WSDs.

7.2.5.2 Enforcement

With WSD hardware readily available, hackers may hack into the WSD, intercept WSDB queries and transmit using protected frequencies illegally (rouge transmission). While this is not happening yet, the regulator or any enforcement body must be ready for such incident to happen. To crack down on such rouge use of frequencies, high-sensitivity tracking is required. The tracking could be done using multiple monitoring towers with high-gain antennas and high-sensitivity receivers, mobile scanners, or via collection of big data from WSDB and analyze for irregular patterns.

7.3 APPLICATIONS AND BUSINESS MODEL

History showed us repeatedly that we need an inflexion point in order for a new technology to take off, for example, Wi-Fi started to take off in a big way after Intel announced they were going to put a Wi-Fi chip in every laptop. For TVWS to take off in a big way, a killer application or innovative business model is required.

In recent years, sharing economy is booming. From Airbnb to Uber, underutilized resources are shared to address user needs. These businesses are very successful since they address a gap between supply and demand. Similarly, shared spectrum access is having the same concept.

Comparing to 3G/4G and Wi-Fi, TVWS does not have certainty of spectrum (except for Singapore TVWS regulation). As a result, businesses find it hard to come up with a convincing business model. With HPCs, businesses will have better guarantee of spectrum availability to address their needs. However, the current HPC model incurs a cost for accessing HPCs. This cost is likely to fluctuate depending on supply and demand. With cost fluctuations, it is hard for businesses to determine the right financial projection. This problem could be solved if the cost of accessing spectrum is very low compared to the total cost of providing services.

Most applications of TVWS focus on providing connectivity for long-range communications or IoT. In fact, the concept of using WSDB provides an additional opportunity for new businesses. With data collected by WSDB, many potential applications are possible. In fact, WSDB becomes a "sensor-less sensor network."

A sensor network usually has a large number of sensors. Deploying and maintaining these sensors will incur a high cost. When WSDs are widely used, WSDs could act as sensors without causing inconvenience to the WSD users. A WSDB could provide service to WSDs with minimal or without charges. In return, the WSDB could request WSDs to feedback radio environment sensing information back to WSDB. If the number of WSDs is large, the WSDB could collect sufficient data for processing. With these data, various services such as detection of change in landscape, coarse positioning, anomaly detection, and so on are possible.

The WSDB operator can either operate the sensor networks on its own or alternatively sell the data to another sensor network operator. This *sensorless sensor network* is powerful as the potential applications are only limited by our imagination.

7.4 SUMMARY

In summary, TVWS has some initial successes and is gaining momentum globally. It is the beginning of a whole new paradigm of spectrum sharing via DSA.

Spectrum sharing has the potential to improve spectrum utilization many folds due to the current inefficient way of using spectrum. With dynamic sharing in nature, it also has the potential to adapt to the ever changing regulatory, business, technical, and environment landscapes.

The WSDB method adopted for TVWS is generally accepted by the community. It can also be used to manage licensed bands or together with other licensed exempt bands. This platform also has the potential to become the "digital regulator" where regulations can be changed and adapted on the fly, allowing regulators to make bolder moves in regulations.

TVWS proves to be a strong candidate in addressing the needs especially for rural broadband, fiber/cable replacement as well as IoT. It has the potential of addressing applications that were previously not possible. Based on the fundamentals and potentials described in this book, TVWS and DSA are here to stay!

REFERENCES

1. Wireless Innovation Forum , Dynamic Spectrum Sharing Annual Report: 2014, Document WINNF-14-P-0001, August 2014.
2. Electronic Communications Committee (ECC), European Conference of Postal and Telecommunications Administrations (CEPT) , Refarming and Secondary Trading in a Changing Radiocommunications World, Report 16, September 2002.
3. D. Bharadia, E. McMilin, and S. Katti, *Full Duplex Radios*, ACM Sigcomm, 2013.
4. X. Zhang and E. W. Knightly, *WATCH: Wi-Fi in Active TV Channels*, ACM MobiHoc, 2015.

Appendix A

Dynamic Spectrum Alliance Model White Spaces Rules[*]

The model white spaces rules that follow are designed to be a template, on which to base rules for license–exempt use of TV white spaces. They are based on the existing U.S. and proposed UK regulations, to ensure the benefits of globally harmonized rules and access to globally standard technologies.

To be sure, the process of providing license–exempt spectrum access to white space devices (WSDs) will vary across different jurisdictions. Accordingly, the model rules also incorporate some flexibility to diverse legal environments and regulatory regimes. For example, in some cases, regulators can enable white space devices by a simple amendment to existing rules, while in other cases these changes may require legislative changes. However, the underlying technical approaches identified in the model rules for enabling white space device operation can be applied in the vast majority of jurisdictions.

One critical concern is whether regulators can permit license–exempt access to the TV white space spectrum under the International Telecommunications Union's (ITU) radio regulations. The definitive answer is "yes." The ITU World Radiocommunication Conference of 2012 concluded that the current international regulatory framework can accommodate software-defined radio and cognitive radio systems, hence dynamic spectrum access, without being changed. According to Francois Rancy, Director of the ITU's Radiocommunications Bureau, "The development of systems implementing this concept, such as TV white spaces, is therefore essentially in the hands of national regulators in each country."

Nonetheless, ITU member states regulators could benefit from collaborative approaches with other member states. Several studies currently underway in ITU-R working parties (WPs), namely, 1B, 5A, 5D, and 6A, will serve as the basis for discussions about TV white spaces and dynamic spectrum access at WRC-15. ITU-R WP5A has recently finalized a technical report on cognitive radio systems.

[*] *Source:* Reproduced with permission from *Suggested Technical Rules and Regulations for the Use of Television White Spaces*, Dynamic Spectrum Alliance Limited. http://www.dynamicspectrumalliance.org

TV White Space: The First Step Towards Better Utilization of Frequency Spectrum, First Edition.
Ser Wah Oh, Yugang Ma, Ming-Hung Tao, and Edward Peh.
© 2016 by The Institute of Electrical and Electronics Engineers, Inc. Published 2016 by John Wiley & Sons, Inc.

Moreover, the ITU development (ITU-D) sector is taking on a greater role in assessing the social and economic benefits of spectrum sharing approaches, including dynamic spectrum access. In particular, the World Telecommunications Development Conference (WTDC) in 2015 provided a formal mandate – Resolution 9 – for further study on dynamic spectrum access within the upcoming ITU-D study cycle. The next study cycle of ITU-D will thus offer further opportunities for ITU Member states to share their experiences on implementing regulations and to demonstrate the impact of TV white spaces in their developmental efforts.

In summary, the model rules include provisions that provide a technical framework to enable license–exempt access to valuable, underutilized spectrum while protecting incumbent licensees. Under the model rules:

- The rules are designed to enable additional access to spectrum *while protecting incumbents from harmful interference*. A primary goal of the rules is to ensure that incumbent operations are not disrupted.
- License–exempt white space devices can use one of two separate methods to prevent harmful interference to incumbents: geolocation or spectrum sensing.
- The geolocation method requires white space devices to determine their physical location and to avoid incumbent licensees in their vicinity by contacting a database that contains information about incumbents and the frequencies, on which they operate. This method is designed with a fail-safe; white space devices may operate only after receiving permission from a database, and may be shut down if there is a risk of harmful interference.
- The regulator may designate one public entity or multiple private entities to administer databases. The rules also describe database administrator responsibilities.
- White space devices relying on a database must communicate with that database in a secure manner. A white space device must cease operations if a database indicates that the frequencies in use by the device are no longer available. Incorporation of a database enables greater control over devices, thereby providing additional assurance that devices will not interfere with incumbent operations.
- Regardless of the mechanism that a white space device uses to avoid causing harmful interference, all devices must comply with operational rules, such as transmit power and emissions limits, to protect incumbents.

§ 1 *Permissible Frequencies of Operation*

    (a) White space devices ("WSDs") are permitted to operate on a license–exempt basis subject to the interference protection requirements set forth in these rules.

    (b) WSDs may operate in the broadcast television frequency bands, as well as any other frequency bands designated by [Regulator].

    (c) WSDs shall only operate on available frequencies determined in accordance with the interference avoidance mechanisms set forth in §2.

(**d**) Client WSDs shall only operate on available frequencies determined by the database and provided via a master white space device in accordance with §3(f).

§ 2 *Protection of Licensed Incumbent Services*

Availability of frequencies for use by WSDs may be determined based on the geolocation and database method described in §3 or based on the spectrum sensing method described in §6.1.

§ 3 *Geolocation and Database Access*

(**a**) A WSD may rely on the geolocation and database access mechanism described in this section to identify available frequencies.

(**b**) WSD geolocation determination.

(**1**) The geographic coordinates of a fixed WSD shall be determined to an accuracy of ±50 m by either automated geolocation or a professional installer. The geographic coordinates of a fixed WSD shall be determined at the time of installation and first activation from a power-off condition, and this information shall be stored by the device. If the fixed WSD is moved to another location or if its stored coordinates become altered, the operator shall reestablish the device's geographic location either by means of automated geolocation or through the services of a professional installer.

(**2**) A personal/portable master WSD shall use automated geolocation to determine its location. The device shall report its geographic coordinates as well as the accuracy of its geolocation capability (e.g., +/−50 m, +/− 100 m) to the database. A personal/portable master device must also reestablish its position each time it is activated from a power-off condition and use its geolocation capability to check its location at least once every 60 s while in operation, except while in sleep mode, that is, a mode in which the device is inactive but not powered down.

(**c**) Determination of available frequencies and maximum transmit power.

(**1**) Master WSDs shall access a geolocation database designated by [Regulator] over the Internet to determine the frequencies and maximum transmit power available at the device's geographic coordinates. A database will determine available frequencies and maximum transmit power based on the algorithm described in §4. However, in no case shall the maximum transmit power exceed the values provided in §7.

(**2**) Master devices must provide the database with the device's geographic coordinates in WGS84 format, a unique alphanumeric code supplied by the manufacturer that identifies the make and model of the device, and a unique device identifier such as the serial number. Fixed master devices must also provide the database with the antenna height of the transmitting antenna specified in meters above ground level or above mean sea level.

(3) When determining frequencies of operation and maximum transmit power, the geolocation database may also take into account additional information voluntarily provided by a master WSD about its operating parameters and indicate to the WSD that different frequencies and/or higher maximum transmit power are available based on this additional information.

(4) WSD operation in a frequency range must cease transmitting immediately if the database indicates that the frequencies are no longer available.

(5) A personal/portable master device must access a geolocation database as described in paragraph (c)(1) to recheck the database for available frequencies and maximum operating power when (1) the device changes location by more than 100 m from the location, at which it last accessed the database or (2) the device is activated from a power-off condition.

(6) A personal/portable master WSD may load frequency availability information for multiple locations around, that is, in the vicinity of, its current location and use that information in its operation. A personal/portable master WSD may use such available frequency information to define a geographic area, within which it can operate on the same available frequencies at all locations; for example, a master WSD could calculate a bounded area, in which frequencies are available at all locations within the area and operate on a mobile basis within that area. A master WSD using such frequency availability information for multiple locations must contact the database again if/when it moves beyond the boundary of the area where the frequency availability data are valid, and must access the database daily even if it has not moved beyond that range to verify that the operating frequencies continue to be available. Operation must cease immediately if the database indicates that the frequencies are no longer available.

(d) Time validity and database recheck requirements. A geolocation database shall provide master devices with a time period of validity for the frequencies of operation and maximum transmit power values described in paragraph (c).

(e) Fixed device registration.

(1) Prior to operating for the first time or after changing location, a fixed WSD must register with a database by providing the information listed in paragraph (e)(3) of this section.

(2) The party responsible for a fixed WSD must ensure that a database has the most current, up-to-date information for that device.

(3) The database shall contain the following information for fixed WSDs:

(i) A unique alphanumeric code supplied by the manufacturer that identifies the make and model of the device (in jurisdictions that require a certification ID number this ID number may be used)

(ii) Manufacturer's serial number of the device

(iii) Device's geographic coordinates (latitude and longitude (WSG84)

(iv) Device's antenna height above ground level or above mean sea level (meters)

(v) Name of the individual or business that owns the device

(vi) Name of a contact person responsible for the device's operation.

(vii) Address of the contact person

(viii) Email address of the contact person

(ix) Phone number of the contact person

(f) Client device operation.

(1) A client WSD may only transmit upon receiving a list of available frequencies and power limits from a master WSD that has contacted a database. To initiate contact with a master device, a client device may transmit on available frequencies used by the master WSD or on frequencies that the master WSD indicates are available for use by a client device on a signal seeking such contacts. A client WSD may optionally provide additional information about its operating parameters to a master device that may be taken into account by the database when determining available frequencies and/or maximum transmit power for the client device. The client device must also provide the master device with a unique alphanumeric code supplied by the manufacturer that identifies the make and model of the client device, which will be supplied to a geolocation database.

(2) At least once every 60 s, except when in sleep mode, that is, a mode in which the device is inactive but is not powered down, a client device must communicate with a master device, which may include contacting the master device to reverify/reestablish frequency availability or receiving a contact verification signal from the master device that provided its current list of available frequencies. A client device must cease operation immediately if it has not communicated with the master device as described above after more than 60 s. In addition, a client device must recheck/reestablish contact with a master device to obtain a list of available frequencies if the client device resumes operation from a powered-down state. If a master device loses power and obtains a new frequency list, it must signal all client devices it is serving to acquire a new frequency list.

(g) Fixed devices without a direct connection to the Internet. If a fixed WSD does not have a direct connection to the Internet and has not yet been initialized and communicated with a geolocation database consistent with this section, but can receive the transmissions of a master

WSD, the fixed WSD needing initialization may transmit to the master WSD either on a frequency band on which the master WSD has transmitted or on a frequency band that the master WSD indicates is available for use to access the geolocation database to receive a list of frequencies and power levels that are available for the fixed WSD to use. Fixed devices needing initialization must transmit at the power levels specified under the technical requirements in these rules for the applicable frequency bands. After communicating with the database, the fixed WSD must then only use the frequencies and power levels that the database indicates are available for it to use.

(**h**) Security.

    (**1**) For purposes of obtaining a list of available frequencies and related matters, master WSDs shall be capable of contacting only those geolocation databases operated by administrators authorized by [Regulator].

    (**2**) Communications between WSDs and geolocation databases are to be transmitted using secure methods that ensure against corruption or unauthorized modification of the data; this requirement also applies to communications of frequency availability and other spectrum access information between master devices.

    (**3**) Communications between a client device and a master device for purposes of obtaining a list of available frequencies shall employ secure methods that ensure against corruption or unauthorized modification of the data. Contact verification signals transmitted for client devices are to be encoded with encryption to secure the identity of the transmitting device. Client devices using contact verification signals shall accept as valid for authorization only the signals of the device from which they obtained their list of available frequencies.

    (**4**) Geolocation database(s) shall be protected from unauthorized data input or alteration of stored data. To provide this protection, a database administrator shall establish communications authentication procedures that allow master devices to be assured that the data they receive are from an authorized source.

§ **4** *Database Algorithm.*

    (**a**) The input to a geolocation database will be positional information from a master WSD, a classification code or other information characterizing a device's emissions performance,[1] the height of the transmitting

---

[1] The European Telecommunications Standards Institute (ETSI) defines five different classes of emissions masks. If available, this information should be supplied to the database. If not, the device can provide its emissions performance to the database in another form. If a device is sophisticated enough to modify its emissions profile dynamically, then regulators can consider an approach in which the database provides a maximum power level per channel and then the device ensures—based on its emissions profile—that it falls below the ceiling provided by the database.

antenna for fixed master devices, and use by licensed incumbents in or near the geographic area of operation of the WSD. The database may, at its discretion, accept additional information about WSD operating parameters. The database will supply a list of available frequencies and associated radiated powers to WSDs pursuant to either: (1) the algorithm provided in Sections A.1 and A.2 or (2) the algorithm provided in Section A.4. Section A.3 provides guidance for implementing either algorithm.[2]

(b) Information about incumbent licensed usage typically will be provided from information contained in [Regulator's] databases.

(c) Any facilities that [Regulator] determines are entitled to protection but not contained in [Regulator's] databases shall be permitted to register with a geolocation database pursuant to §5.

§ 5 *Database Administrator*

(a) Database administrator responsibilities. [Regulator] will designate one public entity or multiple private entities to administer geolocation database(s). Each geolocation database administrator designated by [Regulator] shall fulfill the following:

(1) Maintain a database that contains information about incumbent licensees to be protected.

(2) Implement propagation algorithms and interference parameters issued by [Regulator] pursuant to §4 to calculate operating parameters for WSDs at a given location. Alternatively, a database operator may implement other algorithms and interference parameters that can be shown to return results that provide at least the same protection to licensed incumbents as those supplied by [Regulator]. Database operators will update the algorithms or parameter values that have been supplied by [Regulator] after receiving notification from [Regulator] that they are to do so.

(3) Establish a process for acquiring and storing in the database necessary and appropriate information from the [Regulator's] databases and synchronizing the database with current [Regulator] databases at least once a week to include newly licensed facilities or any changes to licensed facilities.

(4) Establish a process for the database administrator to register fixed WSDs.

(5) Establish a process for the database administrator to include in the geolocation database any facilities that [Regulator] determines are

---

[2] The DSA supports models that protect incumbents but maximize spectrum utility. To that end, they support models that use point-to-point modeling. In addition, they support models that take into account the variability in terrain in calculating propagation and spectrum availability. Sections A.1 and A.2 describe the model that meets these criteria. It is based on the Longley–Rice propagation model. However, the DSA believes that ITU-R.P-1812 is also an acceptable propagation model for this purpose. Details regarding ITU-R.P-1812 are set forth in Sections A.4. Other models may also be appropriate, provided that they use point-to-point calculations and take into account terrain variability.

entitled to protection but not contained in a database maintained by [Regulator].

(**6**) Provide accurate information regarding permissible frequencies of operation and maximum transmit power available at a master WSD's geographic coordinates based on the information provided by the device pursuant to §3(c). Database operators may allow prospective operators of WSDs to query the database and determine whether there are vacant frequencies at a particular location.

(**7**) Establish protocols and procedures to ensure that all communications and interactions between the database and WSDs are accurate and secure and that unauthorized parties cannot access or alter the database or the list of available frequencies sent to a WSD.

(**8**) Respond in a timely manner to verify, correct and/or remove, as appropriate, data in the event that [Regulator] or a party brings a claim of inaccuracies in the database to its attention. This requirement applies only to information that [Regulator] requires to be stored in the database.

(**9**) Transfer its database, along with a list of registered fixed WSDs, to another designated entity in the event it does not continue as the database administrator at the end of its term. It may charge a reasonable price for such conveyance.

(**10**) The database must have functionality such that upon request from [Regulator] it can indicate that no frequencies are available when queried by a specific WSD or model of WSDs.

(**11**) If more than one database is developed for a particular frequency band, the database administrators for that band shall cooperate to develop a standardized process for providing on a daily basis or more often, as appropriate, the data collected for the facilities listed in subparagraph (5) to all other WSD databases to ensure consistency in the records of protected facilities.

(**b**) Nondiscrimination and administration fees.

(**1**) Geolocation databases must not discriminate between devices in providing the minimum information levels. However, they may provide additional information to certain classes of devices.

(**2**) A database administrator may charge a fee for provision of lists of available frequencies to fixed and personal/portable WSDs (and for registering fixed WSDs).

(**3**) [Regulator], upon request, will review the fees and can require changes in those fees if they are found to be excessive.

§ 6 *Spectrum Sensing in the Broadcast Television Frequency Bands*

(**a**) Parties may submit applications for authorization of WSDs that rely on spectrum sensing to identify available frequencies in the television broadcast bands. WSDs authorized under this section must demonstrate that they will not cause harmful interference to incumbent licensees in those bands.

(b) Applications shall submit a preproduction WSD that is electrically identical to the WSD expected to be marketed, along with a full explanation of how the WSD will protect incumbent licensees against harmful interference. Applicants may request that commercially sensitive portions of an application be treated as confidential.

(c) Application process and determination of operating parameters.

(1) Upon receipt of an application submitted under this section, [Regulator] will develop proposed test procedures and methodologies for the preproduction WSD. [Regulator] will make the application and proposed test plan available for public review, and afford the public an opportunity to comment.

(2) [Regulator] will conduct laboratory and field tests of the preproduction WSD. This testing will be conducted to evaluate proof of performance of the WSD, including characterization of its sensing capability and its interference potential. The testing will be open to the public.

(3) Subsequent to the completion of testing, [Regulator] will issue a test report, including recommendations for operating parameters described in subparagraph (c)(4), and afford the public an opportunity to comment.

(4) After completion of testing and a reasonable period for public comment, [Regulator] shall determine operating parameters for the production WSD, including maximum transmit power and minimum sensing detection thresholds, that are sufficient to enable the WSD to reliably avoid harmfully interfering with incumbent services.[3]

(d) Other sensing requirements. All WSDs that rely on spectrum sensing must implement the following additional requirements:

(1) *Frequency availability check time.* A WSD may start operating on a frequency band if no incumbent licensee device signals above the detection threshold determined in subparagraph (c) are detected within a minimum time interval of 30 s.

(2) *In-service monitoring.* A WSD must perform in-service monitoring of the frequencies used by the WSD at least once every 60 s. There is no minimum frequency availability check time for in-service monitoring.

(3) *Frequency move time.* After an incumbent licensee device signal is detected on a frequency range used by the WSD, all transmissions by the WSD must cease within 2 s.

§ 7 *Technical Requirements for WSDs Operating in the Television Broadcast Bands*

(a) Maximum power levels.

(1) WSDs relying on the geolocation and database method of determining channel availability may transmit using the power levels

---

[3] In the context of television broadcast services, the partners suggest that harmfully interfering with an *otherwise viewable* television signal would not be permitted under these guidelines.

provided by the database pursuant to §4. However, the maximum conducted power delivered to the antenna system for WSDs shall never exceed the following values:

(i) The maximum conducted power delivered to the antenna system shall not exceed 16.2 dBm/100 kHz[4,5]. If transmitting antennas of directional gain of greater than 6 dBi are used, this conducted power level shall be reduced by the amount, in dB, that the directional gain of the antenna exceeds 6 dBi.

(ii) Personal/portable WSDs devices shall be treated the same as fixed devices, except[6] under the following conditions:

   a. If the personal/portable WSD does not report its height information, it will be treated like a fixed devices operating at 1.5 m above ground.

   b. If the personal/portable WSD does report its height information, and that height is more than 2 m above ground, an additional 7 dB of power may be permitted beyond what is allowed for fixed devices.

(iii) Fixed WSDs communicating with a master WSD for the purpose of establishing initial contact with a geolocation database pursuant to §3(g) may transmit using the maximum power levels in this paragraph applicable to personal/portable WSDs.

(2) WSDs relying on the spectrum sensing method of determining channel availability may transmit at 50 mW per [television channel size] and −0.4 dBm/100KHz effective isotropic radiated power (EIRP).

§ 8 *Definitions*

(a) *Available frequency.* A frequency range that is not being used by an authorized incumbent service at or near the same geographic location as the WSD and is acceptable for use by a license–exempt device under the provisions of this subpart. Such frequencies are also known as white space frequencies (WSFs).

(b) *Client device.* A personal/portable WSD that does not use an automatic geolocation capability and access to a geolocation database to obtain a list of available frequencies. A client device must obtain a list of available frequencies, on which it may operate from a master device. A client device may not initiate a network of fixed and/or personal/portable WSDs nor

---

[4] A trial in Cape Town, South Africa, in which several DSA members participated, operated at 4 W immediately adjacent to broadcast operations, and no interference was detected. The power level recommended in these rules corresponds to a maximum of 10 W effective isotropic radiated power (EIRP) in a 6 MHz channel. In actual deployments, the power is likely to be limited further by incumbent operation and the device's emissions profile.

[5] The calculation of maximum conducted power under this rule should take into account the transmit power of the radio as well as the loss from cable and connectors.

[6] According to Ofcom's proposed technical rules, portable devices located more than 2 m above ground are presumed to be indoor. The additional power adjustment accounts for building loss.

may it provide a list of available frequencies to another client device for operation by such device.

(c) *Contact verification signal.* An encoded signal broadcast by a master device for reception by client devices to which the master device has provided a list of available frequencies for operation. Such signal is for the purpose of establishing that the client device is still within the reception range of the master device for purposes of validating the list of available frequencies used by the client device and shall be encoded to ensure that the signal originates from the device that provided the list of available frequencies. A client device may respond only to a contact verification signal from the master device that provided the list of available frequencies, on which it operates. A master device shall provide the information needed by a client device to decode the contact verification signal at the same time it provides the list of available frequencies.

(d) *Fixed device.* A WSD that transmits and/or receives radiocommunication signals at a specified fixed location. A fixed WSD may select frequencies for operation itself from a list of available frequencies provided by a geolocation database and initiate and operate a network by sending enabling signals to one or more fixed WSD and/or personal/portable WSDs.

(e) *Geolocation capability.* The capability of a WSD to determine its geographic coordinates in WGS84 format. This capability is used with a geolocation database approved by the [Regulator] to determine the availability of frequencies at a WSD's location.

(f) *Master device.* A fixed or personal/portable WSD that uses a geolocation capability and access to a geolocation database, either through a direct connection to the Internet or through an indirect connection to the Internet by connecting to another master device, to obtain a list of available frequencies. A master device may select a frequency range from the list of available frequencies and initiate and operate as part of a network of WSDs, transmitting to and receiving from one or more WSD. A master device may also enable client devices to access available frequencies by (1) querying a database to obtain relevant information and then serving as a database proxy for the client devices with which it communicates; or (2) relaying information between a client device and a database to provide a list of available frequencies to the client device.

(g) *Network initiation.* The process by which a master device sends control signals to one or more WSDs and allows them to begin communications.

(h) *Operating frequency.* An available frequency used by a WSD for transmission and/or reception.

(i) *Personal/portable device.* A WSD that transmits and/or receives radiocommunication signals at unspecified locations that may change.

(j) *Sensing only device.* A WSD that uses spectrum sensing to determine a list of available frequencies.

(k) *Spectrum sensing.* A process whereby a WSD monitors a frequency range to detect whether frequencies are occupied by a radio signal or signals from authorized services.

(l) *White space device.* An intentional radiator that operates on a license–exempt basis on available frequencies.

(m) *Geolocation database.* A database system that maintains records of all authorized services in the frequency bands approved for WSD use, is capable of determining available frequencies at a specific geographic location, and provides lists of available frequencies to WSDs. Geolocation databases that provide lists of available frequencies to WSDs must be authorized by [Regulator].

A.1 GENERALIZED DESCRIPTION OF PROPAGATION MODEL

A.1.1 Introduction

The Model Rules for the Use of Television White Spaces contemplate that available frequencies and maximum transmit power for a white space device at a given location may be determined based on a geolocation and database method.[7] In particular, database(s) designated by the regulator will provide this information based on the positional information from a master white space device, the height of the transmitting antenna (for fixed master devices), and use by licensed incumbents in or near the geographic area of operation of the white space device.[8] A database will supply a list of available frequencies and associated permitted transmit powers to white space devices pursuant to the procedures in Sections A.1, A.2, and A.3[9] or in Sections A.3 and A.4.

These procedures set forth in this document rely on the Longley–Rice radio propagation model, also known as the Irregular Terrain Model ("Longley–Rice" or "ITM"), which predicts median transmission loss over irregular terrain relative to free space transmission loss.[10] Section A.1 (this section) provides a generalized description of the algorithm used by the Longley–Rice model; Section A.2 describes the elements to be taken into account when implementing the Longley–Rice methodology for television broadcasting service to obtain television broadcasting station field strength values at a particular geographic location; and Section A.3 sets forth the method by which a database operator uses the relevant inputs to indicate available frequencies and maximum power limits for white space devices.

---

[7] See model rules for license–exempt white space devices in § 3 ("Model Rules").

[8] Id. § 4.

[9] Id.

[10] See the U.S. Department of Commerce, National Telecommunications & Information Administration, Institute for Telecommunication Sciences, Irregular Terrain Model (ITM) (Longley–Rice) (20 MHz to 20 GHz), at http://www.its.bldrdoc.gov/resources/radio-propagation-software/itm/itm.aspx.

A.1.2 The Longley–Rice Algorithm

The Longley–Rice model is specifically intended for computer use. The Institute for Telecommunication Sciences ("ITS"), a research and engineering laboratory of the National Telecommunications and Information Administration ("NTIA") within the U.S. Department of Commerce, maintains the "definitive" representation of the Longley–Rice model, which is written in the FORTRAN computing language.[11] In addition, ITS provides a detailed description of the algorithm used by the Longley–Rice model.[12]

Because this document is widely referenced, this section reproduces below much of the original text of the algorithm description provided by ITS, including the original numerical identifiers for sections and equations.

A.1.2.1 Input

The Longley–Rice model includes two modes—the *area prediction mode* and the *point-to-point mode–* which are distinguished mostly by the amount of input data required. The point-to-point mode must provide details of the terrain profile of the link that the area prediction mode will estimate using empirical medians. Since in other respects the two modes follow very similar paths, the ITS algorithm description addresses both modes in parallel.

General Input for Both Modes of Usage

d	Distance between the two terminals
h_{g1}, h_{g2}	Antenna structural heights
k	Wave number, measured in units of reciprocal lengths; see Note 1
Δh	Terrain irregularity parameter
N_s	Minimum monthly mean surface refractivity, measured in N-units; see Note 2
γ_e	The earth's effective curvature, measured in units of reciprocal length; see Note 3
Z_g	Surface transfer impedance of the ground–a complex, dimensionless number; see Note 4
Radio climate	Expressed qualitatively as one of a number of discrete climate types

Note 1: The wave number is that of the carrier or central frequency. It is defined to be

$$k = 2\pi/\lambda = f/f_0 \quad \text{with} \quad f_0 = 47.70\,\text{MHz m}, \tag{A.1}$$

where λ is the wavelength and f is the frequency. Here and elsewhere we have assumed the speed of light in air is 299.7 m/μs.

---

[11] See id.

[12] See generally George Hufford, *The ITS Irregular Terrain Model*, version 1.2.2, the Algorithm (1995). Available at http://www.its.bldrdoc.gov/media/35878/itm_alg.pdf.

Note 2: To simplify its representation, the surface refractivity is sometimes given in terms of N_0, and the surface refractivity "reduced to sea level." When this is the situation, one must know the general elevation z_s of the region involved, and then

$$N_s = N_0 e^{-z_s/z_1}, \quad \text{with} \quad z_1 = 9.46 \text{ km}. \tag{A.2}$$

Note 3: The earth's effective curvature is the reciprocal of the earth's effective radius and may be expressed as

$$\gamma_e = \gamma_a/K,$$

where γ_a is the earth's actual curvature and K is the "effective earth radius factor." The value is normally determined from the surface refractivity using the empirical formula

$$\gamma_e = \gamma_a \left(1 - 0.04665 e^{N_s/N_1}\right), \tag{A.3}$$

where

$$N_1 = 179.3 \, N\text{-units} \quad \text{and} \quad \gamma_a = 157 \times 10^{-9} \text{ m}^{-1} = 157 \, N\text{-units/km}.$$

Note 4: The "surface transfer impedance" is normally defined in terms of the relative permittivity ε_r and conductivity σ of the ground, and the polarization of the radio waves involved. In these terms, we have

$$Z_g = \begin{cases} \sqrt{\varepsilon_r' - 1}, & \text{horizontal polarization}, \\ \dfrac{\sqrt{\varepsilon_r' - 1}}{\varepsilon_r'}, & \text{vertical polarization}, \end{cases} \tag{A.4}$$

where ε_r' is the "complex relative permittivity" defined by

$$\varepsilon_r' = \varepsilon_r + iZ_0\sigma/k, \quad Z_0 = 376.62 \, \Omega. \tag{A.5}$$

The conductivity σ is normally expressed in siemens (reciprocal ohms) per meter.

A.1.2.1.1 Additional Input for the Area Prediction Mode

Siting criteria   Criteria describing the care taken at each terminal to ensure good radio propagation conditions. This is expressed qualitatively in three steps: at random, with care, and with great care

A.1.2.1.2 Additional Input for the Point-to-Point Mode

h_{e1}, h_{e2} Antenna effective heights
d_{L1}, d_{L2} Distances from each terminal to its radio horizon
θ_{e1}, θ_{e2} Elevation angles of the horizons from each terminal at the height of the antennas. These are measured in radians

These quantities, together with Δh, are all geometric and should be determined from the terrain profile that lies between the two terminals. We shall not go into detail here.

The "effective height" of an antenna is its height above an "effective reflecting plane" or above the "intermediate foreground" between the antenna and its horizon. A difficulty with the model is that there is no explicit definition of this quantity, and the accuracy of the model sometimes depends on the skill of the user in estimating values for these effective heights.

In the case of a line-of-sight path, there are no horizons, but the model still requires values for $d_{Lj}, \theta_{ej}, j = 1, 2$. They should be determined from the formulas used in the area prediction mode and listed in Section A.1.2.3. Now it may happen that after these computations, one discovers $d > d_L = d_{L1} + d_{L2}$, implying that the path is a beyond- horizon one. Noting that d_L is a monotone increasing function of the h_{ej}, we can assume these latter have been underestimated and that they should be increased by a common factor until $d_L = d$.

A.1.2.2 Output

The output from the model may take on one of several forms at the user's option. Simplest of these forms is just the *reference attenuation A_{ref}*. This is the *median* attenuation relative to a free space signal that should be observed on the set of all similar paths during times when the atmospheric conditions correspond to a standard, well-mixed atmosphere.

The second form of output provides the two- or three-dimensional cumulative distribution of attenuation in which time, location, and situation variability are all accounted for. This is done by giving the *quantile A(qT,qL,qS)*, the attenuation that will not be exceeded as a function of the fractions of time, locations, and situations. One says *In qS of the situations there will be* at least *qL of the locations where the attenuation* does not exceed *A(qT,qL,qS) for* at least *qT of the time*.

When the point-to-point mode is used on particular, well-defined paths with definitely fixed terminals, there is no location variability, and one must use a two-dimensional description of cumulative distributions. One can now say *With probability (or confidence) qS, the attenuation will not exceed A(qT, qS) for at least qT of the time*. The same effect can be achieved by setting $qL = 0.5$ in the three-dimensional formulation.

On some occasions it will be desirable to go beyond the three-dimensional quantiles and to treat directly the underlying model of variability. For example, consider the case of a communications link that is to be used once and once only. For such a "one-shot" system, one is interested only in what probability or confidence an adequate signal is received that once. The three-dimensional distributions used above must now be combined into one.

A.1.2.3 Preparatory Calculations

We start with some preliminary calculations of a geometric nature.

Preparatory Calculations for the Area Prediction Mode

The parameters $h_{ej}, d_{Lj}, \theta_{ej}, j = 1, 2$, which are part of the input in the point-to-point mode, are, in the area prediction mode, estimated using empirical formulas in which Δh plays an important role.

First, consider the effective heights. This is where the siting criteria are used. We have

$$h_{ej} = h_{gj}, \quad \text{if terminal } j \text{ is sited at random.} \tag{A.6}$$

Otherwise, let

$$B_j = \begin{cases} 5 \text{ m}, & \text{if terminal } j \text{ is sited with care,} \\ 10 \text{ m}, & \text{if termial } j \text{ is sited with great care.} \end{cases}$$

Then

$$B'_j = (B_j - H_1) \sin\left(\frac{\pi}{2} \min\left(\frac{h_{g1}}{H_2}, 1\right)\right) + H_1, \quad \text{with} \quad H_1 = 1m, \quad H_2 = 5m,$$

and

$$h_{ej} = h_{gj} + B'_j e^{-2h_{gj}/h}. \tag{A.7}$$

The remaining parameters are quickly determined:

$$d_{Lsj} = \sqrt{2h_{ej}/\gamma_e},$$

$$d_{Lj} = d_{Lsj} \exp\left[-0.07\sqrt{h/\max(h_{ej}, H_3)}\right], \quad \text{with} \quad H_3 = 5m, \tag{A.8}$$

and, finally,

$$\theta_{ej} = \left[0.65h\left(\frac{d_{Lsj}}{d_{Lj}} - 1\right) - 2h_{ej}\right]/d_{Lsj}. \tag{A.9}$$

Preparatory Calculations for Both Modes

$$d_{Lsj} = \sqrt{2h_{ej}/\gamma_e}, \quad j = 1, 2, \tag{A.10}$$

$$d_{Ls} = d_{Ls1} + d_{Ls2}, \tag{A.11}$$

$$d_L = d_{L1} + d_{L2}, \tag{A.12}$$

$$\theta_e = \max(\theta_{e1} + \theta_{e2}, -d_L\gamma_e). \tag{A.13}$$

We also note here the definitions of two functions of a distance s:

$$\Delta h(s) = \left(1 - 0.8e^{-(s/D)}\right)h, \quad \text{with} \quad D = 50 \text{ km}, \tag{A.14}$$

and

$$\sigma_h(s) = 0.78\Delta h(s) \exp\left[-(\Delta h(s)/H)^{1/4}\right], \quad \text{with} \quad H = 16 \text{ m}. \tag{A.15}$$

A.1.2.4 The Reference Attenuation

The reference attenuation is determined as a function of the distance d from the piecewise formula:

$$A_{\text{ref}} = \begin{cases} \max\left(0, A_{\text{el}} + K_1 d + K_2 \ln\left(d/d_{\text{Ls}}\right)\right), & d \le d_{\text{Ls}}, \\ A_{\text{ed}} + m_d d, & d_{\text{Ls}} \le d \le d_x, \\ A_{\text{es}} + m_s d, & d_x \le d, \end{cases} \qquad (A.16)$$

where the coefficients A_{el}, K_1, K_2, A_{ed}, m_d, A_{es}, m_s, and the distance d_x are calculated using the algorithms below. The three intervals defined here are called the line-of-sight, diffraction, and scatter regions, respectively. The function in (A.16) is continuous so that at the two endpoints where $d = d_{\text{Ls}}$ or d_x the two formulas give the same results. It follows that instead of seven independent coefficients, there are really only five.

A.1.2.4.1 Coefficients for the Diffraction Range Set

$$X_{ae} = \left(k\gamma^2 e\right)^{-1/3}, \qquad (A.17)$$

$$d_3 = \max\left(d_{\text{Ls}}, d_{\text{L}} + 1.3787 X_{ae}\right), \qquad (A.18)$$

$$d_4 = d_3 + 2.7574 X_{ae}, \qquad (A.19)$$

$$A_3 = A_{\text{diff}}(d_3), \qquad (A.20)$$

$$A_4 = A_{\text{diff}}(d_4), \qquad (A.21)$$

where A_{diff} is the function defined below. The formula for A_{ref} in the diffraction range is then just the linear function having the values A_3 and A_4 at the distances d_3 and d_4, respectively. Thus,

$$m_d = (A_4 - A_3)/(d_4 - d_3), \qquad (A.22)$$

$$A_{\text{ed}} = A_3 - m_d d_3. \qquad (A.23)$$

The Function $A_{\text{diff}}(s)$
We first define the weighting factor

$$\omega = \frac{1}{1 + 0.1\sqrt{Q}}, \qquad (A.24)$$

with

$$Q = \min\left(\frac{k}{2\pi}\Delta h(s), 1000\right)\left(\frac{h_{\text{e1}} h_{\text{e2}} + C}{h_{\text{g1}} h_{\text{g2}} + C}\right)^{1/2} + \frac{d_{\text{L}} + \theta_{\text{e}}/\gamma_{\text{e}}}{s} \qquad (A.25)$$

and

$$C = \begin{cases} 0, & \text{in the area prediction mode,} \\ 10\,\text{m}^2, & \text{in the point-to-point mode,} \end{cases}$$

and where $\Delta h(s)$ is the function defined in (A.14) above. Next we define a "clutter factor":

$$A_{f0} = \min\left[15, 5 \log\left(1 + \alpha k h_{g1} h_{g2} \sigma_h(d_{Ls})\right)\right], \quad \text{with} \quad \alpha = 4.77 \times 10^{-4} \, \text{m}^{-2} \quad \text{(A.26)}$$

and with $\sigma h(s)$ defined in (A.15) above.

Then,

$$A_{\text{diff}}(s) = (1 - w)A_k + wA_r + Af_0, \quad \text{(A.27)}$$

where the "double knife edge attenuation" A_k and the "rounded earth attenuation" A_r are yet to be defined. Set

$$\theta = \theta_e + s\gamma e \quad \text{(A.28)}$$

$$v_j = \frac{\theta}{2}\left(\frac{k}{\pi}\frac{d_{Lj}(s - d_L)}{s - d_L + d_{Lj}}\right)^{1/2}, \quad j = 1, 2, \quad \text{(A.29)}$$

and then

$$A_k = F_n(v1) + F_n(v2), \quad \text{(A.30)}$$

where $F_n(v)$ is the Fresnel integral defined below.

For the rounded earth attenuation, we use a "three radii" method applied to Volger's formulation of the solution to the smooth, spherical earth problem. We set

$$\gamma 0 = \theta/(s - d_L), \quad \gamma_j = 2hej/d^2Lj, \quad j = 1, 2, \quad \text{(A.31)}$$

$$\alpha_j = (k/\gamma_j)^{1/3}, \quad j = 0, 1, 2, \quad \text{(A.32)}$$

$$K_j = \frac{1}{i\alpha_j Z_g}, \quad j = 0, 1, 2, \quad \text{(A.33)}$$

Note that the K_j are complex numbers. To continue, we set

$$x_j = AB(K_j)\alpha_j\gamma_j d_{Lj}, \quad j = 1.2, \quad \text{(A.34)}$$

$$x_0 = AB(K_0)\alpha_0\theta + x_1 + x_2 \quad \text{(A.35)}$$

and then

$$A_r = G(x_0) - F(x_1, K_1) - F(x_2, K_2) - C_1(K_0), \quad \text{(A.36)}$$

where $A = 151.03$ is a dimensionless constant and the functions $B(K), G(x), F(x, K)$, and $C1(K)$ are those defined by Vogler.

In (A.30) and (A.36) we have finished the definition of A_{diff}. We should, however, like to complete the subject by defining more precisely the more or less standard functions mentioned above. The Fresnel integral, for example, may be written as

$$Fn(v) = 20 \log\left|\frac{1}{\sqrt{2i}}\int_v^\infty e^{i\pi u^2/2}du\right|. \quad \text{(A.37)}$$

For Vogler's formulation to the solution to the spherical earth problem, we first introduce the special Airy function:

$$Wi(z) = Ai(z) + iBi(z)$$
$$= 2Ai\left(e^{2\pi i/3}z\right),$$

where $Ai(z)$ and $Bi(z)$ are the two standard Airy functions defined in many texts. They are analytic in the entire complex plane and are particular solutions to the differential equation:

$$w"(z) - zw(z) = 0.$$

First, to define the function $B(K)$, we find the smallest solution to the modal equation:

$$Wi(t_0) = 2^{1/3}KW_i'(t_0)$$

and then

$$B = 2^{-1/3} \text{ Im } \{t_0\}. \tag{A.38}$$

Finally, we also have

$$G(x) = 20 \log (x^{-1/2}e^{x/A}), \tag{A.39}$$

$$F(x, K) = 20 \log \left|(\pi/(2^{1/3}AB))^{1/2}Wi(t_0 - (x/(2^{1/3}AB))^2)\right|, \tag{A.40}$$

$$C_1(K) = 20 \log \left|\frac{1}{2}(\pi/(2^{1/3}AB))^{1/2}(2^{2/3}K^2t_0 - 1)Wi'(t_0)^2, \right| \tag{A.41}$$

where A is again the constant defined above.

It is of interest to note that for large x we find $F(x, K) \sim G(x)$, and that for those values of K in which we are interested, it is a good approximation to say $C_1(K) = 20 \text{ dB}$.

A.1.2.4.2 Coefficients for the Line-of-Sight Range We begin by setting

$$d_2 = d_{\text{Ls}} \tag{A.42}$$

$$A_2 = A_{\text{ed}} + m_d d_2. \tag{A.43}$$

Then there are two general cases. First, if $A_{\text{ed}} \geq 0$,

$$d_0 = \min \left(\frac{1}{2}d_{\text{L}}, 1.908kh_{\text{e1}}h_{\text{e2}}\right), \tag{A.44}$$

$$d_1 = \frac{3}{4}d_0 + \frac{1}{4}d_{\text{L}}, \tag{A.45}$$

$$A_0 = A \text{ los } (d_0), \tag{A.46}$$

$$A_1 = A \text{ los } (d_1), \tag{A.47}$$

where the function $Alos(s)$ is defined below. The idea, now, is to devise a curve of the form

$$A_{\text{el}} + K_{1d} + K_2 \ln (d/d_{\text{Ls}}),$$

that passes through the three values A_0, A_1, A_2 at d_0, d_1, d_2, respectively. In doing this, however, we require K_1, $K_2 \geq 0$, and sometimes this forces us to abandon one or both of the values A_0, A_1. we first define

$$K'_2 = \max \left[0, \frac{(d_2 - d_0)(A_1 - A_0) - (d_1 - d_0)(A_2 - A_0)}{(d_2 - d_0) \ln (d_1/d_0) - (d_1 - d_0) \ln (d_2/d_0)} \right], \tag{A.48}$$

$$K'_1 = (A_2 - A_0 - K'2 \ln (d_2/d_0))/(d_2 - d_0), \tag{A.49}$$

which, except for the possibility that the first calculation for K'_2 results in a negative value, is simply the straightforward solution for the two corresponding coefficients. If $K'_1 \geq 0$, we then have

$$K_1 = K'_1, \quad K_2 = K'_2. \tag{A.50}$$

If, however, $K'_1 < 0$, we define

$$K''_2 = (A_2 - A_0)/\ln (d_2/d_0) \tag{A.51}$$

and if now $K''_2 \geq 0$, then

$$K_1 = 0, \quad K_2 = K''_2. \tag{A.52}$$

Otherwise, we abandon both A_0 and A_1 and set

$$K_1 = m_d, \quad K_2 = 0. \tag{A.53}$$

In the second general case we have $A_{ed} < 0$. We then set

$$d_0 = 1.908kh_{e1}h_{e2} \tag{A.54}$$

$$d_1 = \max(-A_{ed}/md, d_L/4). \tag{A.55}$$

If $d_0 < d_1$. we again evaluate A_0, A_1, and K'_2 as before. If $K'_2 > 0$, we also evaluate K'_1 and proceed exactly as before. If, however, we have either $d_0 \geq d_1$ or $K'_2 = 0$, we evaluate A_1 and define

$$K''_1 = (A_2 - A_1)/(d_2 - d_1). \tag{A.56}$$

If now $K''_1 = 0$, we set

$$K_1 = K''_1, \quad K_2 = 0. \tag{A.57}$$

Otherwise, we use (A.53).

At this point, we will define the coefficients K_1 and K_2. We finally set

$$A_{el} = A_2 - K_1 d_2. \tag{A.58}$$

The Function Alos(s)
First, we define the weighting factor:

$$w = 1/\left(1 + \frac{D_1 k \Delta h}{\max (D_2, \, d_{Ls})}\right), \quad \text{with} \quad D_1 = 47.7\,m, \quad D_2 = 10\,\text{km}. \tag{A.59}$$

Then,

$$A_{los} = (1 - w)A_d + wA_t, \tag{A.60}$$

where the "extended diffraction attenuation" A_d and the "two-ray attenuation" A_t are yet to be defined.

First, the extended diffraction attenuation is given very simply by

$$A_d = A_{ed} + m_{ds}. \tag{A.61}$$

For the two-ray attenuation, we set

$$\sin \psi = \frac{h_{e1} + h_{e2}}{\sqrt{s^2 + (h_{e1} + h_{e2})^2}} \tag{A.62}$$

and

$$R'_e = \frac{\sin \psi - Z_g}{\sin \psi + Z_g} \exp\left[-k\sigma_h(s) \sin \psi\right], \tag{A.63}$$

where $\sigma h(s)$ is the function defined in (A.15) above. Note that R'_e is complex since it uses the complex surface transfer impedance Z_g. Then,

$$R_e = \begin{cases} R'_e, & \text{if } |R'_e| \geq \max\left(1/2, \sqrt{\sin \psi}\right), \\ (R'_e/|R'_e|)\sqrt{\sin \psi}, & \text{otherwise.} \end{cases} \tag{A.64}$$

We also set

$$\delta = 2khe1he2/s \tag{A.65}$$

and

$$\delta = \begin{cases} \delta', & \text{if } \delta' \leq \pi/2, \\ \pi - (\pi/2)^2/\delta', & \text{otherwise.} \end{cases} \tag{A.66}$$

Then, finally,

$$A_t = -20 \log |1 + R_e e^{i\delta}|. \tag{A.67}$$

Coefficients for the Scatter Range Set

$$d_5 = d_L + D_s, \tag{A.68}$$

$$d_6 = d_5 + D_s, \quad \text{with} \quad D_s = 200 \text{ km}. \tag{A.69}$$

Then define

$$A_5 = A_{scat}(d_5), \tag{A.70}$$

$$A_6 = A_{scat}(d_6), \tag{A.71}$$

where $A_{scat}(s)$ is defined below. There are, however, some sets of parameters for which A_{scat} is not defined, and it may happen that either or both $A5$, $A6$ are undefined. If this is so, one merely sets

$$dx = +\infty \tag{A.72}$$

and one can let A_{es}, m_s remain undefined. In the more normal situation, one has

$$m_s = (A_6 - A_5)/D_s, \tag{A.73}$$

$$dx = \max \left[d_{Ls}, d_L + X_{ae} \log (kH_s), (A_5 - A_{ed} - m_s d_5)/(m_d - m_s) \right], \tag{A.74}$$

$$A_{es} = A_{ed} + (m_d - m_s)dx, \tag{A.75}$$

where D_s is the distance given above, X_{ae} has been defined in (A.17), and $H_s = 47.7$ m.

The Function As_{cat}

Computation of this function uses an abbreviated version of the methods described in Section 9 and Annex III.5 of NBS TN101.[13] First, set

$$\theta = \theta_e + \gamma_e s, \tag{A.76}$$

$$\theta = \theta_{e1} + \theta_{e2} + \gamma_e s, \tag{A.77}$$

$$r_j = 2k\theta' h_{ej}, \quad j = 1, 2. \tag{A.78}$$

If both r_1 and r_2 are less than 0.2, the function A_{scat} is not defined (or is infinite). Otherwise, we put

$$A_{scat}(s) = 10 \log \left(kH\theta^4 \right) + F(\theta_s, N_s) + H_0, \tag{A.79}$$

where $F(\theta_s, N_s)$ is the function shown in Figure 9.1 of TN101, H_0 is the "frequency gain function," and $H = 47.7$ m.

The frequency gain function H_0 is a function of r_1, r_2, the scatter efficiency factor η_s, and the "asymmetry factor" that we shall here call ss. A difficulty with the present model is that there is not sufficient geometric data in the input variables to determine where the crossover point is. This is resolved by assuming it to be midway between the two horizons. The asymmetry factor, for example, is found by first defining the distance between horizons:

$$d_s = s - d_{L1} - d_{L2}, \tag{A.80}$$

whereupon

$$S_s = \frac{d_{L2} + d_s/2}{d_{L1} + d_s/2}. \tag{A.81}$$

There then follows that the height of the crossover point is

$$z_0 = \frac{S_s d\theta'}{(1 + S_s)^2} \tag{A.82}$$

---

[13] See P. L. Rice, A. G. Longley, K. A. Norton, and A. P. Barsis, *Transmission loss predictions for tropospheric communication circuits*, U.S. Government Printing Office, Washington, DC, NBS Tech. Note 101, issued May 1965; revised May 1966 and January 1967 ("TN101").

and then

$$\eta_s = \frac{Z_0}{Z_0}\left[1 + (0.031 - N_s 2.32 \times 10^{-3} + N_s^2 5.67 \times 10^{-6})e^{-(z_0/Z_1)^6}\right], \qquad \text{(A.83)}$$

where

$$Z_0 = 1.756\,\text{km}, \quad Z_1 = 8.0\,\text{km}.$$

The computation of H_0 then proceeds according to the rules in Section 9.3 and Figure 9.3 of TN101.

The model requires these results at the two distances $s = d_5, d_6$, described above. One further precaution is taken to prevent anomalous results. If, at d_5, calculations show that H_0 will exceed 15 dB, they are replaced by the value it has at d_6. This helps keep the scatter-mode slope within reasonable bounds.

A.1.2.5 Variability: The Quantiles of Attenuation

We want now to compute the quantiles $A(q_T, q_L, q_S)$ where q_T, q_L, q_S, are the desired fractions of time, locations, and situations, respectively. In the point-to-point mode, we would want a twofold quantile $A(q_T, q_S)$, but in the present model this is done simply by computing the three-fold quantile with q_L equal to 0.5.

Because the distributions involved are all normal, or nearly normal, it simplifies the calculations to rescale the desired fractions and to express them in terms of "standard normal deviates." We use the complementary normal distribution

$$Q(z) = \frac{1}{\sqrt{2\pi}} \int_z^\infty e^{-t^2/2} dt$$

and then the deviate is simply the inverse function

$$z(q) = Q^{-1}(q).$$

Thus, if the random variable x is normally distributed with mean X_0 and standard deviation σ, its quantiles are given by

$$X(q) = X0 + \sigma z(q).$$

Setting

$$z_T = z(q_T), \quad z_L = z(q_L), \quad z_S = z(q_S),$$

we now ask for the quantiles $A(z_T, z_L, z_S)$. In these rescaled variables, it is as though all probabilities are to be plotted on normal probability paper. In the case of the point-to-point mode, we will simply suppose $z_L = 0$.

First, we define

$$A' = A_{\text{ref}} - V_{\text{med}} - Y_T - Y_L - Y_S, \qquad \text{(A.84)}$$

where A_{ref} is the reference attenuation defined in Section 4, and where the adjustment V_{med} and the deviations Y_T, Y_L, Y_S are defined below. The values of Y_T and Y_L

depend on the single variables z_T and z_L, respectively. The value of Y_S, on the other hand, depends on all three standard normal deviates.

The final quantile is a modification of A' given by

$$A(z_T, z_L, z_S) = \begin{cases} A', & \text{if } A' \geq 0, \\ A' \dfrac{29 - A'}{29 - 10A'}, & \text{otherwise.} \end{cases} \tag{5.2}$$

An important quantity used below is the "effective distance." We set

$$d_{ex} = \sqrt{2a_1 h_{e1}} + \sqrt{2a_1 h_{e2}} + a_1(kD_1)^{-1/3}, \tag{A.85}$$

with

$$a_1 = 9000 \text{ km}, \quad D_1 = 1266 \text{ km}.$$

Then the effective distance is given by

$$d_e = \begin{cases} D_0 d/d_{ex}, & \text{for } d \leq d_{ex}, \\ D_0 + d - d_{ex}, & \text{for } d \geq d_{ex}, \end{cases} \tag{A.86}$$

with $D_0 = 130$ km.

A.1.2.5.1 Time Variability

Quantiles of time variability are computed using a variation of the methods described in Section 10 and Annex III.7 of NBS TN101, and also in CCIR Report 238-3. Those methods speak of eight or nine discrete radio climates, of which seven have been documented with corresponding empirical curves. It is these empirical curves to which we refer below. They are all curves of quantiles of deviations versus the effective distance d_e.

The adjustment from the reference attenuation to the all-year median is

$$V_{med} = V_{med}(d_e, c \lim) \tag{A.87}$$

where the function is described in Figure 10.13 of TN101.

The deviation Y_T is piecewise linear in z_T, and may be written in the form

$$Y_T = \begin{cases} \sigma_T - z_T, & z_T \leq 0, \\ \sigma_T + z_T, & 0 \leq z_T \leq z_D, \\ \sigma_T + z_D + \sigma_{TD}(z_T - z_D), & z_D \leq z_T. \end{cases} \tag{A.88}$$

The slopes (or "pseudo-standard deviations")

$$\sigma_{T-} = \sigma_T - (d_e, c \lim) \tag{A.89}$$
$$\sigma_{T+} = \sigma_T + (d_e, c \lim)$$

are obtained from TN101 in the following way. For σ_{T-}, we use the 0.90 quantile and divide the corresponding ordinates by $z(0.90) = -1.282$. For σ_{T+}, we use the 0.10 quantile and divide by $z(0.10) = 1.282$.

Table A.1   Ducting (low probability) Constants

Climate	q_D	z_D	C_D
Equatorial	0.10	1.282	1.224
Continental subtropical	≈0.015	2.161	.801
Maritime subtropical	0.10	1.282	1.380
Desert	0	∞	–
Continental temperate	0.10	1.282	1.224
Maritime temperate overland	0.10	1.282	1.518
Maritime temperate oversea	0.10	1.282	1.518

The remaining constants in (5.6) pertain to the "ducting" or low probability, case. We write

$$z_D = z_D(c\,\lim), \qquad \sigma_{TD} = C_D(c\,\lim)\sigma_T+ \tag{A.90}$$

where values of z_D and C_D are given in Table A.1. In this table we have also listed values of $q_D = Q(z_D)$.

A.1.2.5.2 Location Variability

We set

$$Y_L = \sigma_L z_L, \tag{A.91}$$

where

$$\sigma_L = 10k\Delta h(d)/(k\Delta h(d) + 13)$$

and $\Delta h(s)$ is defined in (3.9) above.

A.1.2.5.3 Situation Variability

Set

$$\sigma S = 5 + 3e^{-de/D}, \tag{A.92}$$

where $D = 100$ km. Then

$$Y_S = \left(\sigma_S^2 + \frac{Y_T^2}{7.8 + z_S^2} + \frac{Y_L^2}{24 + z_S^2} \right)^{1/2} z_S. \tag{A.93}$$

The latter is intended to reveal how the uncertainties become greater in the wings of the distributions.

A.1.2.6 Addenda: Numerical Approximations

Part of the algorithm for the ITM consists in approximations for the standard functions that have been used. In these approximations, computational simplicity has often taken greater priority than accuracy.

The Fresnel integral is used in A1.2.4.1 and is defined in (A.37). We have (for $v > 0$)

$$\text{Fn}(v) \approx \begin{cases} 6.02 + 9.11v - 1.27v^2, & \text{if } v \le 2.40. \\ 12.953 + 20\log v, & \text{otherwise.} \end{cases} \tag{A.94}$$

The functions $B(K)$, $G(x)$, $F(x, K)$, $C_l(K)$, which are used in diffraction around a smooth earth, are also used A.1.2.4.1 and are defined in (A.38) to (A.41). We have

$$B(K) \approx 1.607 - |K|, \tag{A.95}$$

$$G(x) = 0.05751x - 10 \log x, \tag{A.96}$$

$$F(x, K) \approx \begin{cases} F_2(x, K), & \text{if } 0 < x \leq 200, \\ G(x) + 0.0134x e^{-x/200}(F_1(x) - G(x)), & \text{if } 200 < x < 2000, \\ G(x), & \text{if } 2000 \leq x, \end{cases} \tag{A.97}$$

where

$$F_1(x) = 40 \log (\max (x, 1)) - 117, \tag{A.98}$$

$$F_2(x, K) = \begin{cases} F_1(x), & \text{if } |K| < 10^{-5} \quad \text{or} \quad x(-\log|K|)^3 > 450, \\ 2.5 \times 10^{-5} x^2 / |K| + 20 \log|K| - 15, & \text{otherwise.} \end{cases} \tag{A.99}$$

The final approximation here is

$$C_1(K) \approx 20. \tag{A.100}$$

To complete this section, we have the two functions, $F(\theta d)$ and H_0, used for tropospheric scatter. First,

$$F(D, N_s) = F_0(D) - 0.1(N_s - 301)e - D/D_0, \tag{A.101}$$

where

$$D_0 = 40 \,\text{km}$$

and (when D is given in meters)

$$F_0(D) = \begin{cases} 133.4 + 0.332 \times 10^{-3} D - 10 \log D, & \text{for } 0 < D \leq 10 \,\text{km}, \\ 104.6 + 0.212 \times 10^{-3} D - 2.5 \log D, & \text{for } 10 < D \leq 70 \,\text{km}, \\ 71.8 + 0.157 \times 10^{-3} D + 5 \log D, & \text{otherwise.} \end{cases} \tag{A.102}$$

The frequency gain function may be written as

$$H_0 = H_{00}(r_1, r_2, \eta_s) + \Delta H_0, \tag{A.103}$$

where

$$\Delta H_0 = 6(0.6 - \log \eta_s) \log s_s \log r_2 / s_s r_1 \tag{A.104}$$

and where H_{00} is obtained by linear interpolation between its values when η_s is an integer. For $\eta_s = 1, \ldots, 5$, we set

$$H_{00}(r_1, r_2, j) = \frac{1}{2}[H_{01}(r_1, j) + H_{01}(r_2, j)], \tag{A.105}$$

with

$$
H_{01}(r,j) = \begin{cases} 10\log\left(1 + 24r^{-2} + 25r^{-4}\right), & j = 1, \\ 10\log\left(1 + 45r^{-2} + 80r^{-4}\right), & j = 2, \\ 10\log\left(1 + 68r^{-2} + 177r^{-4}\right), & j = 3, \\ 10\log\left(1 + 80r^{-2} + 395r^{-4}\right), & j = 4, \\ 10\log\left(1 + 105r^{-2} + 705r^{-4}\right), & j = 5. \end{cases} \tag{A.106}
$$

For $\eta_s > 5$, we use the value for $\eta_s = 5$ and for $\eta_s = 0$, we suppose

$$
H_{00}(r_1, r_2, 0) = 10\log\left[\left(1 + \frac{\sqrt{2}}{r_1}\right)^2 \left(1 + \frac{\sqrt{2}}{r_2}\right)^2 \frac{r_1 + r_2}{r_1 + r_2 + 2\sqrt{2}}\right]. \tag{A.107}
$$

In all of this, we truncate the values of ss and $q = r_2/\mathrm{ss}r_1$ at 0.1 and 10.

A.2. LONGLEY–RICE PARAMETERS FOR TV BROADCAST FIELD STRENGTH CALCULATIONS

A.2.1 Introduction

Section A.2 provides a description of elements to be taken into account in implementing the Longley–Rice radio propagation model—also known as the Irregular Terrain Model (ITM)—in order to use this model to calculate the field strength of a television broadcasting station signal at a particular geographic location. As described in Section A.1, implementations of the Longley–Rice model occur as programs written in a specific computer language. For example, the Institute for Telecommunication Sciences (ITS), a research and engineering laboratory of the National Telecommunications and Information Administration (NTIA) within the U.S. Department of Commerce, maintains the "definitive" representation of the Longley–Rice model, which is written in FORTRAN.[14]

These software implementations of the Longley–Rice radio propagation model will require several inputs to perform the field strength calculation for broadcast television. Although the specific inputs may depend on the particular software implementation used or developed, required parameters/data generally will fall into four categories:

(**a**) Television broadcasting station parameters

(**b**) Planning factors for television reception

(**c**) Longley–Rice environmental parameters

(**d**) Terrain profile data

In addition, certain path calculations—described below—will need to be taken into account in order to predict television broadcasting field strength at a given location.

[14] See generally *Implementation of the Irregular Terrain Model*, version 1.2.2 (updated August 5, 2002). Available at http://www.its.bldrdoc.gov/media/35869/itm.pdf.

A.2.2 Model Parameters

The Longley–Rice radio propagation model can be implemented in "area" mode or "point-to-point" mode. The point-to-point mode is used to evaluate the predicted strength of a particular television channel at a geographic location where a white space device is present. With the point-to-point mode, field strength at a particular geographic location is determined using path-specific parameters determined from detailed terrain profile data. In addition to the location of the WSD and the WSD antenna height (for fixed WSD deployments), software implementations of the Longley–Rice model will require the following input parameters.

A.2.2.1 Television Broadcast Station Parameters

The Longley–Rice model requires the input of television broadcasting station parameters to be used in propagation calculations. For determining accurate field strength values for television stations, the relevant parameters would be those of licensed television stations of interest for each television channel to be evaluated at the WSD location. A software implementation of the Longley–Rice model thus should be designed to access a database of the following relevant licensed television broadcasting station technical characteristics:

- *Frequency:* The carrier frequency of the transmitted broadcast signal.
- *Effective radiated power (ERP):* W, kW, dBW, dBm.
- *Antenna:* Absent additional information, implementations of the Longley–Rice model will assume the use of an omni-directional antenna. This is only an assumption, however, and the model will account for antenna directionality if supplied.
- *Height:* The height of antenna above the ground (supplied in meters or feet); the effective height for calculations should then be estimated by the software implementation of the model.
- *Polarization:* Horizontal polarization should be denoted.

A.2.2.2 Planning Factors for DTV Reception

The planning factors shown in Table A.2 are assumed to characterize the equipment, including antenna systems, used for home reception of DTV signals. They determine the minimum field strength for DTV reception as a function of frequency band and as a function of the channel in the UHF band. Implementations should assume a 10 m height above ground for the television receiving antenna.[15]

---

[15] See Federal Communications Commission, Office of Engineering and Technology Bulletin No. 69, *Longley–Rice Methodology for Evaluating TV Coverage and Interference at 3* (February 6, 2004). Available at http://transition.fcc.gov/Bureaus/Engineering_Technology/Documents/bulletins/oet69/oet69 .pdf ("OET Bulletin No. 69").

Table A.2 Planning Factors for DTV Reception

Planning Factor	Symbol	Low VHF	High VHF	UHF
Geometric mean frequency (MHz)	F	69	194	615
Dipole factor (dBm–dBu)	K_d	−111.8	−120.8	−130.8
Dipole factor adjustment	K_a	None	None	See below
Thermal noise (dBm)	N_t	−106.2	−106.2	−106.2
Antenna gain (dBd)	G	4	6	10
Download line loss (dB)	L	1	2	4
System noise figure (dB)	N_s	10	10	7
Required carrier-to-noise ratio (dB)	C/N	15	15	15

The following dipole factor adjustment calculation, $K_a = 20 \log [615/(\text{channel mid-frequency in MHz})]$, should be added to K_d to account for the fact that field strength requirements are greater for UHF channels above the geometric mean frequency of the UHF band and smaller for UHF channels below that frequency.

A.2.2.3 Longley–Rice Parameters

In addition to the technical operating characteristics for a given broadcast transmitter and planning factors for television reception, the Longley–Rice model contemplates the use of the following parameters that describe the environment in which the transmitter is operating (or, more precisely, statistics about the environment where the transmitter operates):

- *Surface refractivity N_s^*:* This is the refractivity of the atmosphere, measured in N-units (parts per million), which typically ranges from 250 to 400 N-units. ITS guidance on Longley–Rice model implementation[†] includes the values in Table A.3.

- *Permittivity:* This is the dielectric constant of the ground. ITS guidance includes the values in Table A.4.

- *Conductivity:* Soil conductivity of the ground. ITS guidance includes the values in Table A.4.

- *Climate zone:* This value is entered as per climate codes that correspond with the seven climate categories specified in Table A.2. Together with N_s, the climate serves to characterize the atmosphere and its variability in time.

---

[*] Note that the N-Unit value for surface refractivity is a separate parameter unrelated to the symbols denoting noise in Table 1 above.

[†] See Hufford, G. A., A. G. Longley, and W. A. Kissick (1982)280, A guide to the use of the ITS Irregular Terrain Model in the area prediction mode, NTI301A Report 82-100. (NTIS Order No. PB82-217977).

Table A.3 ITS Values for N_s

Radio Climate	N_s (N-units)
Equatorial (Congo)	360
Continental subtropical (Sudan)	320
Maritime subtropical (West Coast of Africa)	370
Desert (Sahara)	280
Continental temperate	301
Maritime temperate, over land (the United Kingdom and continental west coasts)	320
Maritime temperate, over sea	350

For average atmospheric conditions, use a continental temperate climate and $N_s = 301$ N-units.

- *Variability:* The Longley–Rice model includes the following three kinds of variability:
  - *Location variability (reliability and confidence level):* This value is expressed as a percentage from 0.1% to 99.9%. Location variability accounts for variations in long-term statistics that occur from path to path.
  - *Time variability:* This value is expressed as percentage from 0 to 100%. Time variability accounts for variations of median values of attenuation.
  - *Situation variability:* This value is expressed as a percentage; 50% variability is considered normal for coverage estimations. Situation variability accounts for variations between systems with the same system parameters and environmental conditions.
- *Variability modes:* ITS guidance contemplates the following ways in which the kinds of variability listed above are treated in combination:
  - Broadcast mode: all three kinds of variability are treated separately.
  - Individual mode: situation and location variability are combined; time variability is treated separately.
  - Mobile mode: time and location variability are combined; situation variability is treated separately.
  - Single message mode: all three kinds of variability are combined.

Table A.4 ITS Values for Electrical Ground Constants

	Relative Permittivity	Conductivity (S/m)
Average ground	15	0.005
Poor ground	4	0.001
Good ground	25	0.020
Freshwater	81	0.010
Seawater	81	5.0

For most purposes, use the constants for an average ground.

Table A.5 Longley–Rice Parameter Values for Television Signal Analysis

Longley–Rice Parameter	Value
Surface refractivity in N-units (parts per million)	301.0
Relative permittivity of ground	15.0
Ground conductivity (S/m)	0.005
Climate zone code	5 (continental temperate)
Mode for variability calculations	Broadcast mode

The values listed in Table A.5 have historically been utilized when implementing the Longley–Rice model for television signal analysis,[16] and should be used to calculate the field strength of a television broadcasting station signal at a particular geographic location.

A.2.2.4 Terrain Profile Path Data

The Longley–Rice model may use terrain elevation values to create a detailed profile of a path for analysis by the program. The model was designed to use terrain data at equal increments along a path. Points not at equal increments are ignored. Consequently, field strength values are calculated values out to the last uniformly spaced point on a given radial.

A Longley–Rice implementation may achieve greater precision by utilizing values given by specific terrain datasets collected using empirical measurements. For example, the Shuttle Radar Topography Mission (SRTM) undertaken by the National Aeronautics and Space Administration (NASA) obtained elevation data on a near-global scale in order to create a high-resolution digital topographic database of most of the Earth, providing 3 arc-second ($\sim$ 90 m) resolution data for most of the continents between 60 N and 60 S.[17] In many populated areas, higher resolution sources of terrain data are available.

A.2.3 Path Calculations

The Longley–Rice model uses input parameters to compute geometric parameters related to the propagation path. First, the model determines effective antenna height. Since this is an area prediction model, the radio horizons, for example, are unknown. The model uses the terrain irregularity parameter to estimate radio horizons. The model also computes a reference attenuation, using horizon distances and elevation angles to calculate transmission loss relative to free space.

The Longley–Rice model will treat the terrain that separates the television broadcast station and the white space device location as a random function

---

[16] See OET Bulletin No. 69 at 6.

[17] See generally National Aeronautics and Space Administration, *Shuttle radar topography mission: the mission to map the world*, at http://www2.jpl.nasa.gov/srtm/.

Table A.6   ITS Values for Terrain Irregularity

	Δh (m)
Flat (or smooth water)	0
Plains	30
Hills	90
Mountains	200
Rugged mountains	500

For an average terrain, use $\Delta h = 90$ m.

characterized by Δh. The model uses a signal value Δh to represent the size of the irregularities. Roughly speaking, Δh is the interdecile range of terrain elevations—that is, the total range of elevations after the highest 10% and lowest 10% have been removed. Suggested values for Δh provided by ITS are set forth in Table A.6.

A.2.4 Summary

This section describes input parameters, terrain data, and calculations that must be taken into account in implementing software to build an application to calculate the field strength of a television broadcasting station at a particular location. Determining the field strength of relevant television broadcast stations at a specific location will be used to determine at what level and at what power level a WSD may operate at that location pursuant to the procedure outlined in Section A.3.

A.3. CALCULATION OF AVAILABLE TV WHITE SPACE FREQUENCIES AND POWER LIMITS

A.3.1 Introduction

This section provides the detailed parameters and methodology to calculate the frequencies and maximum power limits for white space devices in such as way as to limit the probability of harmful interference to other services to acceptable levels. The proposed methodology in this section is independent of the radio propagation models that might be used in the calculations described herein. However, it is imperative that point-to-point path-specific statistical radio propagation models that are capable of utilizing digital terrain/elevation models must be used.

A.3.2 Definitions

This section describes the various entities and their relationships with regard to frequency and signal strength calculations. Interference from white space devices is controlled by limiting their radiated power. The following definitions describe an approach for how those power limits can be calculated.

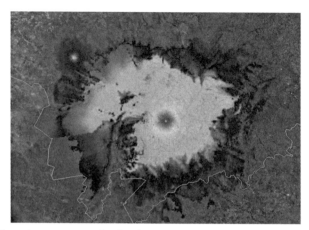

Figure A.1 Example coverage map showing "in-service" areas extending out to blue areas, purple = "too weak" (out of service). (Reproduced with permission of Dynamic Spectrum Alliance Limited.)

A.3.2.1 Protected Services

Analog Terrestrial Television(ATT): PAL-I Standard The service area of an analog TV broadcast includes any locations where the signal to noise ratio of its signal is greater than or equal to 17.0 dB plus a link margin of 13 dB (Figure A.1).[18]

Digital Terrestrial Television (DTT): DVB-T2 Standard The service area of a digital TV broadcast includes any locations where the signal to noise ratio of its signal is greater than or equal to 17.0 dB plus a link margin of 13 dB. Given that the proposed methodology analyzes transmitters individually, single-frequency networks (SFNs) and multi-frequency networks (MFNs) can be treated similarly.

Radio Astronomy Sites (RASs) The protected area of radio astronomy sites (RASs) particularly the square kilometer array (SKA) project sites are designated as radio quiet zones (RQZs). However, this designation also extends to some other SKA sites that are not located within the RQZs. The prescribed maximum received power spectral density (PSD) at each site must not exceed −130 dBm/8 MHz.[19] The methodology for calculating the necessary protection from white space devices is still to be determined and may not be required or applicable in all jurisdictions.

Other Radio astronomy frequencies, lower band edge, upper band edge, wireless microphones/program-making and special events equipment, studio-to-transmitter links, cable TV receive sites and gap filter terrestrial television stations may require additional protection. Further analysis and industry consultation is required. Again, with the exception of upper and lower band edge users, these incumbents may not be present in all jurisdictions.

---

[18] A service's signal-to-noise ratio (SNR) limit is the minimum theoretical operating level for a service to be functional, while the link margin accounts for the extra signal power that is typically required to cope with real-world environments. The link margin provides a buffer so that the service is somewhat robust against common signal impairments like multipath, fading, and interference.

[19] Radio astronomy sensitivity based on ITU-R RA.769-2.

Table A.7 Analog TV Adjacent Channel Rejection Ratios

Channel Offset	Relative Signal Strength (dB)
$N - 10$ or lower	62
$N - 9$	60
$N - 8$	58
$N - 4$	55
$N - 3$	45
$N - 2$	42
$N - 1$	36
$N + 1$	36
$N + 2$	42
$N + 3$	45
$N + 4$	55
$N + 8$	58
$N + 9$	47*
$N + 10$ or higher	62

*The ACR performance on channel $N + 9$ is different than $N - 9$ due to internal tuner design limitations.

A.3.2.2 Receiver Characteristics

Analog Television (PAL-I) An analog TV receiver's sensitivity to white space device interference is a function of its adjacent channel rejection ratio (ACR). If a receiver is tuned to channel "N", it can tolerate signals on adjacent channels without harmful interference if the relative signal strength are less than the values given in Table A.7 (Figure A.2).[20–22]

Digital Television (DVB-T2). A digital TV receiver's sensitivity to white space device interference is a function of its adjacent channel rejection ratio (ACR). If a receiver is tuned to channel "N", it can tolerate signals on adjacent channels without harmful interference if the relative signal strength are less than the values given in Table A.8 (Figure A.3).[23]

Mobile Digital Terrestrial Television (MDTT): DVB-H Standard. The reference minimum media field strength values of DVB-H reception for band IV and band V are prescribed in Table 11.14 and 11.15 in the DVB blue book.[24,25]

---

[20] The ACR performance values are based on an Ofcom analysis of 50 commercially available receivers currently on the market. See Ofcom, TV White Spaces: Approach to Coexistence, Technical Analysis, September 4, 2013. Available at http://stakeholders.ofcom.org.uk/binaries/consultations/white-space-coexistence/annexes/technical-report.pdf (Ofcom Technical Report).

[21] Values between $N \pm 4$ and $N \pm 8$ should be linearly interpolated.

[22] The ACR performance on channel $N + 9$ is different than $N - 9$ due to internal tuner design limitations.

[23] The ACR performance values are based on an Ofcom analysis of 50 commercially available receivers currently on the market. See Ofcom Technical Report.

[24] Values between $N \pm 4$ and $N \pm 8$ should be linearly interpolated.

[25] The ACS performance on channel $N + 9$ is different than $N - 9$ due to internal tuner design limitations.

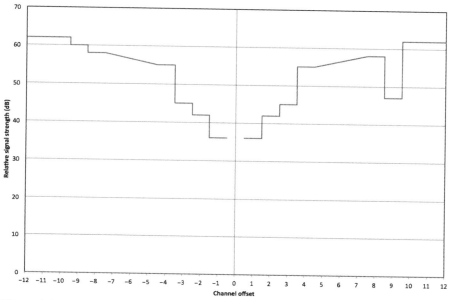

Figure A.2 Plot of analog TV adjacent channel rejection ratios.

A.3.2.3 WSD Coupling Loss

The coupling loss between a white space device and other types of receivers is assumed to be 60 dB.[26]

Table A.8 Digital TV Adjacent Channel Rejection Ratios

Channel Offset	Relative Signal Strength (dB)
$N-10$ or lower	62
$N-9$	60
$N-8$	58
$N-4$	55
$N-3$	45
$N-2$	42
$N-1$	36
$N+1$	36
$N+2$	42
$N+3$	45
$N+4$	55
$N+8$	58
$N+9$	47
$N+10$ or higher	62

[26] The coupling loss accounts for a multitude of factors, including the separation distance between devices, antenna discrimination, polarization discrimination, building attenuation, physical obstructions, and so on.

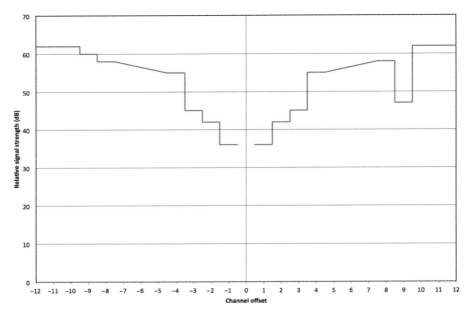

Figure A.3 Plot of digital TV adjacent channel rejection ratios.

A.3.2.4 Resolving Terrain Overlap

Terrain data files are generally organized in "tiles" (rectangular rasters aligned to latitude and longitude bins) that include overlapping data along each of its edges. When overlapping tiles contain nonidentical data in their overlapping zones, there is the potential for elevation ambiguity in those areas.

To resolve this ambiguity, the following tile selection methodology shall be used. For any given point, exactly one terrain tile will be selected as the authoritative source of elevation data.

1. For any given point (`lat` and `lon`), determine the set of tiles that include the requested `lat` and `lon` coordinates. The number of matching tile candidates is expected to be between 0 and 4.
   a. If the number of matching tiles is 0 (point does not fall within any terrain tile), then treat the elevation as 0 m and return.
   b. If the number of matching tiles is 1, then the point does not have any data overlap issues. Use bilinear interpolation to compute the terrain elevation using the selected tile.
2. If 2 or more tile candidates are found, use the following criteria to select which tile to use.
   a. Compute the latitude distance from `lat` to each candidate tile. The tile(s) with the smallest latitudinal distance (`lat_distance`) wins.

b. In case of a tie, compute the longitude distance from `lon` to each candidate tile. The tile(s) with the smallest longitude distance (`lon_distance`) wins.

c. In case of a tie, select the tile with the lowest latitude and lowest longitude coverage (i.e., lowest numerical values).

d. Use bilinear interpolation to compute the terrain elevation using the selected tile.

Hint: This is effectively the same as sorting the candidate tiles according to multiple keys. The primary key is the $lat\_distance_i$, followed by $lon\_distance_i$, $center\_lat_i$, and $center\_lon_i$ to resolve any ties when necessary.

Illustrated Example Consider a sample point that lies in the overlap zone between three tiles.

Note that the tiles do not necessarily have the same raster density or coverage range.

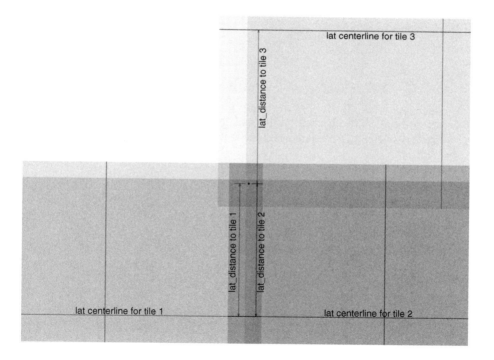

Since there are 2 or more candidate tiles to consider, they need to be ranked according to $lat\_distance_i$.

In this example, Tile 1 and Tile 2 have the same $lat\_distance$. Both tiles are closer to the sample point than Tile 3.

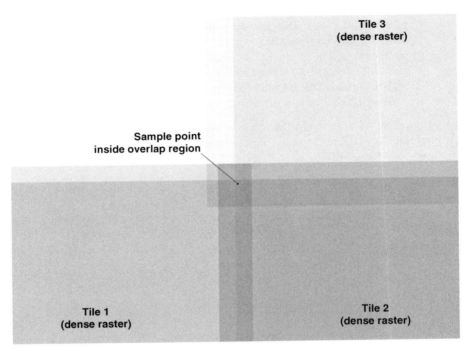

Since the lat_distance for Tile 1 and Tile 2 is the same, the lon_distance needs to be checked. In this example, Tile 2 is closer than Tile 1. Tile 2 is selected as the authoritative tile to use.

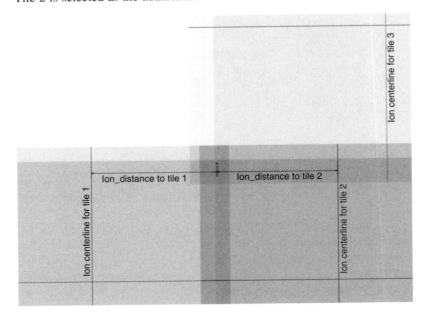

A.3.2.5 Terrain Profile Resolution

The terrain profile between a transmit and receive point is constructed by taking the terrain elevation at equally spaced points along the shortest path between the transmitter and receiver. The shortest path is computed using the Vincenty algorithm.[27] The nominal spacing of the bins in the terrain profile is 50 m. Note that the terrain profile spacing is independent of the underlying terrain grid size. Bilinear interpolation is used to fill points in the terrain profile as described in Section 4 above.

A.3.3 Calculations

White space spectrum availability calculations are location-specific. For the purpose of discussion in this section, the WSD is assumed to be at a point W_0, which has a latitude of $W_{0,lat}$, a longitude of $W_{0,lon}$, and a height of $W_{0,h}$ (optional).

1. Compute frequencies that are "in use" by protected services
   1.1. Identify all of the protected entities that are within 300 km of point W_0.
   1.2. If the height of W_0 is not available, then assume that $W_{0,h} = 10$ m above ground.
   1.3. For each protected entity, apply the point-to-point propagation model prescribed in Sections A.1 and A.2 to predict the residual power level of each entity at point W_0. Example
      1.3.1. For a TV transmitter T_i, compute the effective radiated power (P_i, radiated) of T_i in the direction of W_0, including any antenna pattern adjustments.
      1.3.2. Use propagation modeling to compute the path loss (LT_i) between T_i and W_0.
      1.3.3. Compute the effective ambient signal power at point W_0 as

$$P_{i,\text{eff}} = P_{i,\text{radiated}} - LT_i$$

      1.3.4. If $P_{i,\text{eff}}$ is greater than the SNR limit plus link margin for TV transmitters, then this channel is considered to be "in use" by T_i, otherwise the signal is too weak and this channel is considered to be "not in use" by T_i.

T_i
$P_{i, radiated}$

W_0
$P_{i, eff}$

---

[27] The Vincenty inverse algorithm (http://www.ngs.noaa.gov/PUBS_LIB/inverse.pdf) computes the shortest ellipsoidal path between two points on an oblate spheroid (like the WGS84 reference model of the Earth).

2. For each "in use" protected service, use the adjacent channel rejection ratios relative to $P_{i,\text{eff}}$ to compute the power constraints that should be applied to white space devices.

2.1. Television example

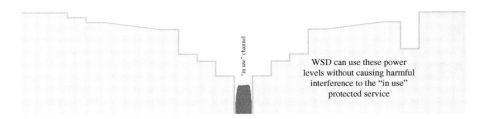

When multiple channels are "in use", their relative adjacent channel rejection ratios can overlap. The adjacent channel restriction with the *lowest* power level is used to determine the power limit for WSD use since this prevents the WSD from causing interference on any of the active services.

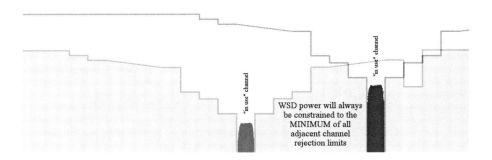

After all protected services have been processed, the remaining power envelope defines the maximum permitted power levels for WSD use.

2.2. Band edge protection is specific to each jurisdiction.

The resulting power envelope after all incumbent protections have been processed is known as the baseline spectrum availability profile. This baseline profile can optionally be processed further to compensate for the WSD emission masks as described in Section 3.1.

3. WSD Emission mask considerations

3.1. Predefined emission classes

Some WSDs will report their radio emissions capabilities in the form of a device type or emission class. If the device reports its emission mask characteristics in the form of a device type or class code, the database can compute adjustments to the baseline spectrum availability profile to match the device's emissions characteristics.

For example, devices using ETSI 301 598 emission class codes must satisfy the following adjacent channel leakage limits:

nth Adjacent Channel Number	Adjacent Channel Leakage Ratio (dB)				
	Class 1	Class 2	Class 3	Class 4	Class 5
$n = \pm 1$	74	74	64	54	43
$n = \pm 2$	79	74	74	64	53
$n \geq 3$ or n <= −3	84	74	84	74	64

The database can adjust the baseline spectrum profile such that a given device's emissions class is guaranteed to never exceed the available spectrum power envelope. The database would find the "best fit" for the device that ensures that all of its emissions (both in-band and adjacent channels) stay below the baseline spectrum availability profile.

As an example, suppose we have a baseline spectrum availability profile that has a mix of high and low power channels as follows (this is just for illustrative purposes):

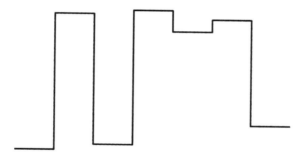

A Class 1 device connecting to the database will have very good adjacent channel leakage performance, so it will be able to emit relatively high power levels and still stay beneath the permitted power envelope.

A Class 5 device will have a worse adjacent channel leakage performance, so it will need to have it's power levels reduced in order to stay within the bounds of the permitted power envelope.

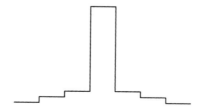

Class 1 emission class.

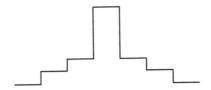

Class 5 emission class.

The database adjusts the spectrum response according to the class of device making the query, and ensures that the devices can never exceed the baseline spectrum profile.

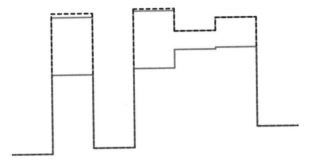

In this example, a Class 1 device will receive a response from the database that is very close to the baseline spectrum profile because it has a very good emission mask that guarantees that it will not exceed the baseline power limits.

For the Class 5 device, the database has to reduce the power on some channels to make sure the device's adjacent channel leakage stays below the baseline power limits. Each device class will receive an answer from the database tailored to its particular emission characteristics.

3.2. Statically defined emission limits

Some WSDs will report their radio emissions capabilities in the form of a device type or mode definition (e.g., fixed, personal/portable modes I and II). If the device reports its characteristics in the form of a device type or mode, the database can compute adjustments to the baseline spectrum availability profile to match the device's emissions characteristics.

For example, devices following the FCC device mode definitions must satisfy the following statically defined emission limits:

Emissions Type	Max EIRP Per 6 MHz	
	Fixed (dBm)	Modes I and II (dBm)
In-band	36	16–20
Out-of-band	−36.8	−56.8

To ensure device compliance, the database can make adjustments to the baseline spectrum availability profile by clamping the power levels to the max in-band power limits corresponding to the device mode. Any channels that are not safe for operation at the in-band power limits will be set to the out-of-band power levels to prevent devices from operating on those channels.

As an example, suppose we have a baseline spectrum availability profile that has a mix of high and low power channels as follows (this is just for illustrative purposes):

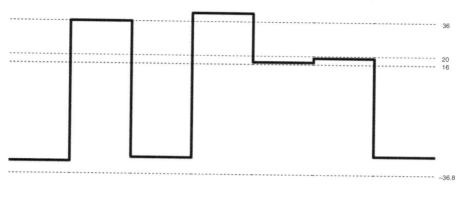

The database adjusts the spectrum response according to the mode of device making the query, and ensures that the device can never exceed the baseline spectrum profile and can never exceed the designated out-of-band emission limits.

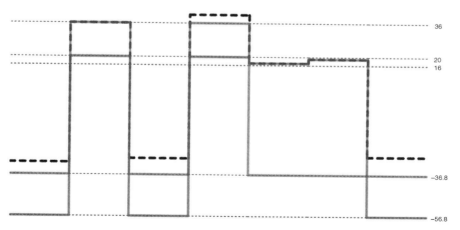

In this example, a fixed device will receive a response from the database that permits operation on any channels that allow for operation at

36 dBm or higher. Any baseline spectrum availability channels that are higher than 36 dBm will be clamped to 36 dBm. All other channels will be set to the out-of-band power limit for fixed devices at −36.8 dBm. This will guarantee that the fixed device will only operate on channels that are safe from interference and that the out-of-band emissions will stay below the defined limits.

In this example, a mode I or II device will receive a response from the database that permits operation on any channels that allow for operation at 16 dBm or higher. Any baseline spectrum availability channels that are higher than 20 dBm will be clamped to 20 dBm. All other channels will be set to the out-of-band power limit for fixed devices at −56.8 dBm. This will guarantee that the mode I or II device will only operate on channels that are safe from interference and that the out-of-band emissions will stay below the defined limits.

3.3. Dynamic emissions control

Some devices may be capable of dynamically controlling their own emissions profile to adapt to the spectrum response provided by the database. If a database is queried by a certified device having such capabilities, the database can respond with the baseline spectrum availability profile, and the device must then ensure that all of its in-band and out-of-band emissions comply with those limits.

When the WSD gets a response from the spectrum database, it must take into consideration its own out-of-band emissions mask to ensure that it will not violate the limits specified in the spectrum response.

A device might have different out-of-band emission characteristics depending on its quality of design (e.g., high-end versus low-end components), or it may use a technology that supports multiple modulation types (e.g., Wi-Fi supports up to 76 different modulation coding schemes), or there might be other factors that cause the emissions profile to change depending on current operating conditions. In each of these situations, it is up to the WSD to ensure that its emissions profile is compliant with the database's spectrum response under all modes of operation (Figure A.4).

The WSD could employ any number of techniques to ensure compliance with the database's spectrum response, including, but not limited to, reducing output power, changing modulation types, switching filters, shrinking occupied bandwidth, and so on.

A.4. INFORMATION REGARDING ITU-R P-1812

A.4.1 Introduction

The Suggested Technical Rules and Regulations for the Use of Television White Spaces contemplate that available frequencies and maximum transmit power for a white space device at a given location may be determined based on a geolocation

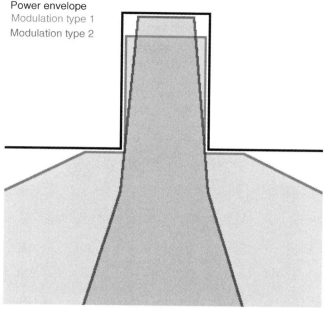

Power envelope
Modulation type 1
Modulation type 2

Figure A.4   Example of WSD emissions profile being considered in conjunction with the power envelope provided by the spectrum database. (Reproduced with permission of Dynamic Spectrum Alliance Limited.)

and database method. In particular, database(s) designated by the regulator will provide this information based on the positional information from a master white space device, the height of the transmitting antenna (for fixed master devices), and use by licensed incumbents in or near the geographic area of operation of the white space device. The rules suggest that a database will supply a list of available frequencies and associated permitted transmit powers to white space devices pursuant to the algorithm in Sections A.1 and A.2 or pursuant to the algorithm set forth in this section. Section A.3 can be used with either Sections A.1 and A.2 or this section.

Sections A.1 and A.2 are based on the Longley–Rice propagation model. However, other point-to-point, terrain-based propagation models may also serve as the basis for spectrum availability calculations. One such model is the International Telecommunication Union's Radiocommunication Sector Recommendation P-1812 (ITU-R. P-1812). Like Longley–Rice, ITU-R. P-1812 is a path-specific propagation prediction method for point-to-area terrestrial services in the VHF and UHF bands. Further detail regarding ITU-R. P-1812, including the model itself and an explanation of its implementation is available for download on the ITU's Web site at https://www.itu.int/rec/R-REC-P.1812-3-201309-I/en.

Appendix B

Performance of SEA*

$\mathbf{F}$igure B.1 shows the return loss comparison of SEA and its original pure antenna. We can see that they have similar impedance matching (S11 < 10 dB). The SEA has impedance-matched frequency range from 490 to 595 MHz, and the original pure antenna's impedance bandwidth is from 490 to 610 MHz.

Figure B.2 shows the radiation pattern of the antennas in the x–y plane. Both the SEA and the original antenna have good radiation uniformity (variation <1 dB) in the x–y plane. The average gain for both SEA and its original version is 1.5 dBi in x–y plane. As for the radiation pattern in the y–z plane, both antennas have the maximum gain in the directions around 0 and −180/180° as shown in Figure B.3.

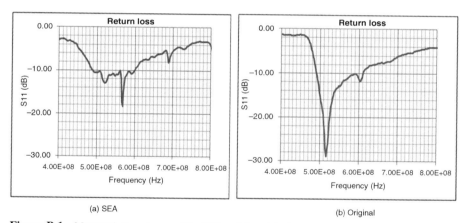

(a) SEA   (b) Original

Figure B.1   Measured return losses of the SEA and the original antenna.

* *Source:* System embedded antenna for portable TVWS cognitive radio transceiver. *IEEE Antennas and Wireless Propagation Letters*, Vol. 14, 2015.

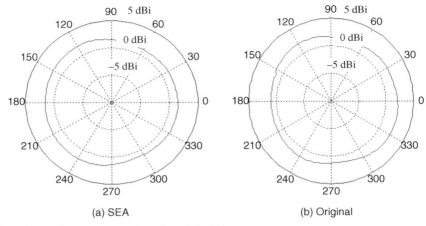

(a) SEA   (b) Original

Figure B.2   Radiation pattern (*x*–*y* plane, 575 MHz).

Figure B.4 shows the radiation in the *x*–*z* plane. Both antennas perform almost the same as that in the *y*–*z* plane. The SEA is an omnidirectional dipole antenna.

The measurement results prove that the SEA has the same properties as the original antenna. In other words, we can design the SEA according to the original antenna instead of considering too much about the surrounding parts and the cross-coupling issue. This makes the system-integrated antenna design more straightforward.

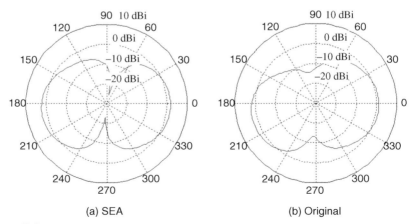

(a) SEA   (b) Original

Figure B.3   Radiation pattern (*y*–*z* plane, 575 MHz).

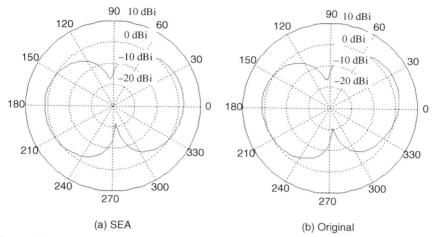

(a) SEA   (b) Original

Figure B.4   Radiation pattern (*x–z* plane, 575 MHz).

Appendix C

Self-Positioning Based on DVB-T2 Signals

A DVB-T2 system is usually deployed as a single-frequency network (SFN) where several transmitters synchronously transmit identical signals over the same frequency channel. To help DVB-T2 network operators in setting up, maintaining, and monitoring the network by identifying the contribution from each individual transmitter, transmitter-dependent signature waveforms are transmitted from every transmitter in the SFN simultaneously to the future extension frame (FEF). FEF, together with the normal DVB-T2 data frames, known as T2 frames, forms the DVB-T2 superframe shown in Figure C.1. The signature waveforms of a FEF shown in Figure C.2 have special cyclic correlation properties, which allow the measurement of channel impulse response and the extraction of timing of the first arrival path from each transmitter. The timing information of multiple transmitters can then be used to compute the location of the positioning WSD.

C.1 DVB-T2-BASED POSITIONING

What we are introducing is a positioning system for WSDs that is more robust to SNR and does not require a prior knowledge of signal propagation loss. In addition, the positioning system should also have the capability to be deployed in large SFNs where the number of transmitters is less than 64.

The basic block diagram for this positioning system is shown in Figure C.3. The received signals, which are a combination of signature waveforms from all transmitters, are first passed to the cyclic correlation block (Figure C.4). The correlation block correlates the received signals with each of the known individual signature waveforms. Due to the zero cyclic cross-correlation among different signature waveforms within the zero-correlation zone, the correlation output for a particular signature waveform consists of only the effective channel response of the transmitter

TV White Space: The First Step Towards Better Utilization of Frequency Spectrum, First Edition.
Ser Wah Oh, Yugang Ma, Ming-Hung Tao, and Edward Peh.
© 2016 by The Institute of Electrical and Electronics Engineers, Inc. Published 2016 by John Wiley & Sons, Inc.

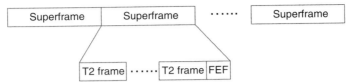

Figure C.1   An example of DVB-T2 frame structure. (*Source*: Digital Video Broadcasting (DVB); Frame Structure Channel Coding and Modulation for a Second-Generation Digital Terrestrial Television Broadcasting System (DVB-T2).)

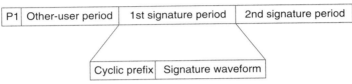

Figure C.2   FEF frame structure (*Source*: Digital Video Broadcasting (DVB); Frame Structure Channel Coding and Modulation for a Second-Generation Digital Terrestrial Television Broadcasting System (DVB-T2).)

Figure C.3   Block diagram of the positioning system of WSDs.

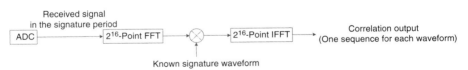

Figure C.4   Block diagram of the cyclic correlation part.

that uses the signature waveform and the noise. The effective channel response is the convolution of the true channel impulse response and the window function due to band-limited operation at the transmitter. Then the correlation output of each signature waveform is passed to the timing estimation block to estimate the timing of the first arrival path (or the first channel tap) of the transmitter that uses the signature waveform. Finally, all the timing estimates are passed to the TDOA location estimation block to estimate the location of the positioning WSD. At least four timing estimates are needed for the TDOA location estimation block to estimate the location.

There are two timing estimation approaches, discussed in the following sections.

C.1.1 Threshold-Based Timing Estimation Approach

To solve the issues described in the literature, a threshold-based timing estimation is adopted. It eliminates the need to perform deconvolution. Instead, it finds the timing that corresponds to the first local maximum above the threshold. In addition, the setting of the threshold only requires the knowledge of noise power but not that of the signal propagation loss. In particular, the threshold is set based on the following guidelines:

(1) The threshold should be set to avoid as much noise as possible.

(2) An extremely erroneous timing estimate can cause large positioning error in the next stage. If the signal from a particular transmitter to the positioning WSD is weak, instead of having an erroneous estimate of the first channel tap, the timing estimate should be avoided to be used for the TDOA location estimation in the next stage, especially in SFNs where there are more than enough transmitters for positioning. In other words, the threshold should have the capability to reject the inaccurate transmitters and ensure only signals from reliable transmitters are used for the TDOA location estimation.

Figure C.5 illustrates the positioning successful rate when position error is within 50 m versus the threshold levels for different received SNR values under

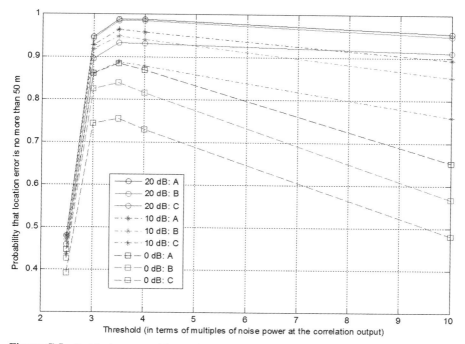

Figure C.5 Positioning successful rate when position error is within 50 m versus the threshold levels for different received SNR values under different channel propagation models.

different channel propagation models. The channel delay profiles of the models A, B, and C are provided in Figures C.12–C.14, respectively. According to the figures, it is suggested that for the received SNR of interest (0–20 dB at the input of the correlation block), the threshold, which is compared with the absolute value of correlation output, is set according to Threshold $= 3.5\sigma^2$, where σ^2 is the noise measured at the output of the correlation block. This corresponds to 10.9 dB signal level above noise floor and it can avoid 99.999% of noise. When the strength of all the channel taps are under this threshold, the transmitter is assumed to be unreliable and is excluded from the TDOA location estimation. The flowchart showing the procedure of the threshold-based timing estimation approach is shown in Figure C.6. An algorithmic implementation of the timing estimation approach is shown in Algorithm C.1.

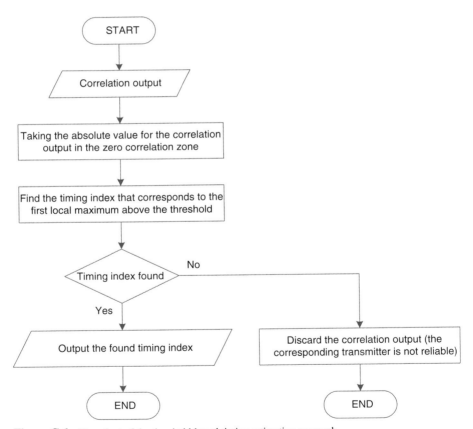

Figure C.6 Flowchart of the threshold-based timing estimation approach.

Algorithm C.1   Threshold-Based Timing Estimation

Input: correlation output $R_{s_jy}[n]$.

Output: Timing estimates for the first channel tap $\hat{n}_j$.

1. Initialize: $n = 0$;

2. While $n \leq \bar{n}$

3.   If correlation value is below the threshold $\left| R_{s_jy}[n] \right| < \text{threshold} \big|$,

4.   If $n = \bar{n}$, then $\hat{n}_j = null$

5.   Else: increment the sample index: $n \leftarrow n + 1$;

6.   Else:

7.   While $n \leq \bar{n}$

8.   If $\left| R_{s_jy}[n + 1] \right| \leq R_{s_jy}[n]$, then: $\hat{n}_j = n$ and exit;

9.   Else: Increment the sample index: $n \leftarrow n + 1$.

C.1.2 Iterative Timing Estimation Approach

The band-limiting operation of the signature waveform at the transmitter causes the spreading of the channel impulse response at the receiver. As a result, adjacent channel taps or channel taps that are close in time may add on to each other constructively or destructively. When the subsequent channel taps that are close in time with the first channel tap are much stronger, using the threshold-based estimation approach described in the preceding section can end up mistaking the timing of the later channel taps with that of the first tap. To solve the problem in such situation, an iterative timing estimation approach is suitable.

The basic principle of the iterative timing estimation approach is as follows: In each iteration multiple channel, taps are estimated from the strongest to the weakest. When estimating each channel taps, the effects of other channel taps are removed based on the latest estimates for these channels. The process continues for a number of iterations until the termination criterion satisfies, for example, the timing estimates are stabilized for a predefined number of iterations. The procedure for the iterative timing estimation approach is shown in Figure C.7. An algorithmic implementation of the above-described iterative timing estimation approach for small SFN is shown in Algorithm C.2.

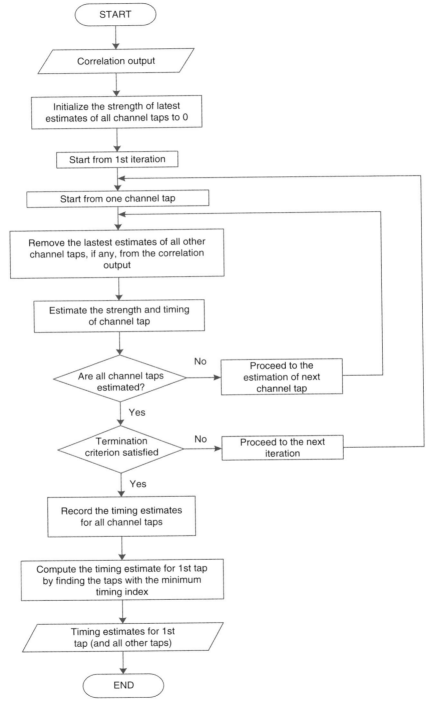

Figure C.7   Flowchart of the iterative timing estimation approach.

Algorithm C.2   Iterative Timing Estimation

Input: correlation output $R_{s_jy}[n]$ and number channel taps L.

Output: timing estimates for the first channel tap $\hat{n}_j$ (and optionally $\hat{n}_{j,l}, l = 1, \ldots, L$).

1. Initialize: $\hat{h}^{(l,d)} = 0$, for $l = 1, \ldots, L, d = 0, 1$; $d = 1$

2. Repeat:

3.   For each channel tap $l \leq L$, do

4.   Cancel the effects of other channel taps from the correlation output based on the latest estimates:

$$R_{s_jy}^{(l,d)}[n] = R_{s_jy}[n] - \sum_{\tilde{l}=1}^{l-1} c\hat{h}^{(\tilde{l}d)} w\left[n - n^{(\tilde{l},d)}\right] - \sum_{\tilde{l}=l+1}^{L} c\hat{h}^{(\tilde{l},d-1)} w\left[n - n^{(\tilde{l},d-1)}\right];$$

5. Estimate the timing and strength of channel tap l: $\hat{n}^{(l,d)} = \arg\max \left| R_{s_jy}^{(l,d)}[n] \right|$ and $\hat{h}^{(l,d)} = \frac{1}{cw[0]} R_{s_jy}^{(l,d)}\left[\hat{n}^{(l,d)}\right]$

6. Increment the iteration index: $d \leftarrow d + 1$;

7. Stop until termination criterion meets and set $\hat{n}_{j,l} = \hat{n}^{(l,d)}$.

8. Compute the timing estimate for the first channel tap by finding the smallest timing among the estimated channel taps: $\hat{n}_j = \min_l \hat{n}_{j,l}$.

Furthermore, for small SFN where the number of transmitter is fewer than eight, each transmitter is uniquely identified by a signature waveform. When the number of transmitter is more than eight, a pair of signature waveforms is used to identify a transmitter: one waveform in the first signature period and the other in the second signature period. Since there are only eight different signature waveforms in total, two transmitters can use the same signature waveforms in one signature period. As a result, the channel impulse response measured after correlation is a mixture of channel responses from two transmitters. A transmitter should be identified by jointly analyzing the channel responses measured in the two signature periods. Threshold-based timing estimation approach can only find the earliest channel tap among two transmitters in each period. It cannot be used to find the first tap of the later arrived transmitter. In contrast, iterative channel estimation approach can identify the timing of all channel taps within each signature period. Then a transmitter can be distinguished from other transmitters by examining the correlation outputs associated with the pair of signature waveforms assigned and finding the common channel taps for the two signature periods. Finally, the time estimates for the first channel tap for a transmitter can be obtained by finding the minimum

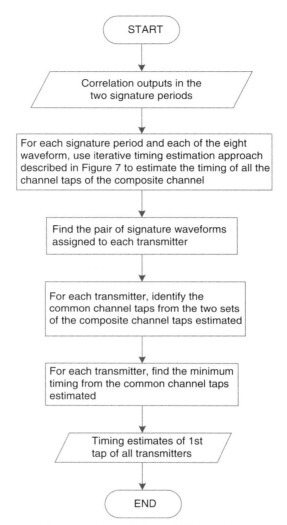

Figure C.8 The process of timing estimation for large SFN where the number of transmitter is between 9 and 64.

timing of the common channel taps. The process of how to find the timing esti-mates for each transmitter is shown in Figure C.8.

In Figures C.9–C.11, some simulation results are presented to show the perform-ance of the positioning system based on the timing estimation approaches. We consider a SFN DVB-T2 network of seven hexagonal cells. Each cell has a radius of 2 km. The positioning WSD is assumed to be uniformly distributed in the central cell. We assume that the propagation loss is given according to $128.1 + 37.6 \log_{10}(d/1 \text{ km}) \text{ dB}$, where d is the distance from the DVB-T2 transmitter to the positioning WSD. We assume log-normal channel shadowing with standard deviation of 8 dB. Three multipath channels

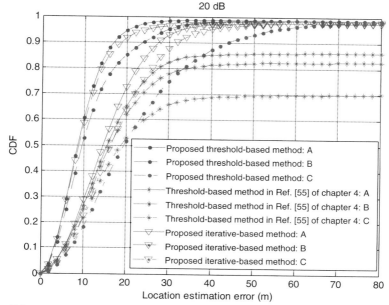

Figure C.9   The cumulative distribution function of the location estimation error for positioning system based on the two timing estimation approaches as compared with the existing approach in Ref. [55] of Chapter 4 under the received SNR of 20 dB.

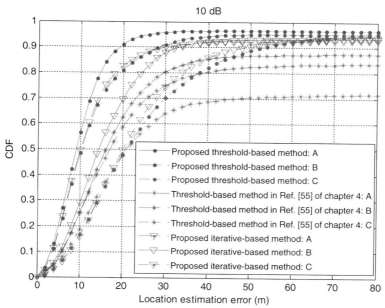

Figure C.10   The cumulative distribution function of the location estimation error for positioning system based on the two timing estimation approaches as compared with the existing approach in Ref. [55] of Chapter 4 under the received SNR of 10 dB.

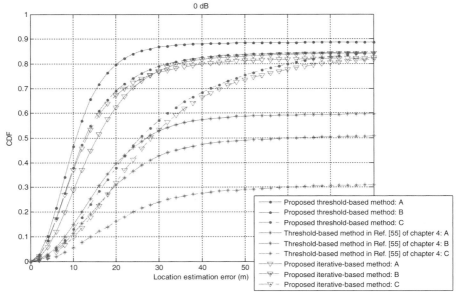

Figure C.11   The cumulative distribution function of the location estimation error for positioning system based on the two timing estimation approaches as compared with the existing approach in Ref. [51] of Chapter 4 under the received SNR of 0 dB.

are considered, whose channel power delay profiles are given in Figures C.12–C.14. This approach is compared with the existing approach in Ref. [55] of Chapter 4 under different received SNR values (20, 10, and 0 dB). It can be seen that the presented schemes exhibit better performance than the existing approach in Ref. [55] of Chapter 4. The performance is improved more significantly as the received SNR reduces. This shows that the presented schemes are more robust to SNR.

This positioning system is also able to give the accuracy of the estimated location in terms of the probability that the location error of the estimated location is within 50 m. This can be done in the following procedures. First, using strength of the estimated channel taps from all transmitters to estimate the total received signal power. Second, estimate the noise power at the correlation output and compute the noise power at the correlation input by multiplying the signal waveform length 2^{16}. Third, compute the estimated SNR value. Finally, based on the cumulative distribution function of the location estimation error for the estimated SNR, find the probability that the location error of the estimated location is within 50 m. If this probability is below the regulatory requirement,[1] the WSD has to wait for the next FEF to estimate the location again.

---

[1] For example, according to the regulation of Ofcom, the accuracy should be within 50 m with a probability of 0.95 (see Ref. [11] of Chapter 4).

Figures C.12–C.14 show the channel power delay profiles for channel models A, B, and C, respectively. These three models are used in evaluating the performance.

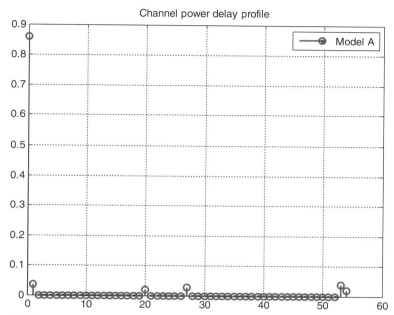

Figure C.12   Channel power delay profile for channel propagation model A.

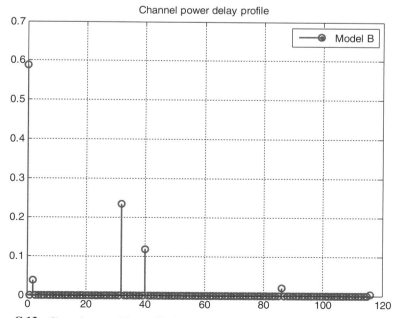

Figure C.13   Channel power delay profile for channel propagation model B.

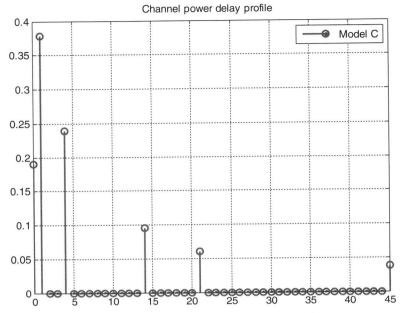

Figure C.14 Channel power delay profile for channel propagation model C.

Appendix D

Algorithm for Dynamic Spectrum Assignment[*]

D.1 SYSTEM MODEL

The system under consideration is a centralized spectrum manager (SM) controlling the access of different WSDs within its coverage area.

We denote the list of available spectrum fragments assigned to the local SM after the query as $\left\{ B_0^{(1)}, \ldots, B_0^{(n)}, \ldots, B_0^{(N)} \right\}$, where $B_0^{(n)}$ represents the bandwidth of the nth spectrum fragment assigned. We consider the case that the requests for bandwidth from different WSDs arrive one by one in a sequential manner. We denote the amount of bandwidth that the tth WSD requested as b_t and the spectrum fragments available at the time of the tth request as $\left\{ B_t^{(1)}, \ldots, B_t^{(n)}, \ldots, B_t^{(N)} \right\}$. The SM knows exactly how much bandwidth is requested only at the time that the request arrives, but does not know *a priori* how much bandwidth is requested for any of the future requests. However, it is assumed that the SM knows the possible set of the requested bandwidth β and their arrival probability distribution. The arrival probability of the different requests can be either independent or follow a Markov transition process. Such information can be easily obtained based on the past observation of the arrival of the requests.

When receiving the request b_t the local SM will decide, based on the current available spectrum fragment $\left\{ B_t^{(1)}, \ldots, B_t^{(n)}, \ldots, B_t^{(N)} \right\}$ and the distribution information about the future requests, if the request can be accepted and if yes, which spectrum fragment n should the request be assigned to. It is considered that different WSDs do not share the same spectrum, that is, the spectrum is assigned to a WSD on an exclusive use basis. It is noted that when the request is accepted, to avoid spectrum

---

[*] *Source:* Dynamic spectrum assignment for white space devices with dynamic and heterogeneous bandwidth requirements. *IEEE Wireless Communications and Networking Conference (WCNC)*, 2015.

TV White Space: The First Step Towards Better Utilization of Frequency Spectrum, First Edition.
Ser Wah Oh, Yugang Ma, Ming-Hung Tao, and Edward Peh.
© 2016 by The Institute of Electrical and Electronics Engineers, Inc. Published 2016 by John Wiley & Sons, Inc.

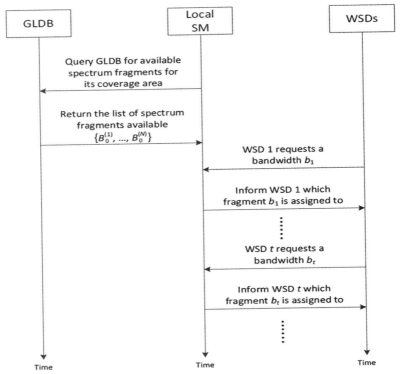

Figure D.1   Illustration of the operation of the TVWS network.

fragmentation the spectrum assigned to b_t is adjacent to edge of the fragment n that is allocated. The process will repeat until there is no more request arrived or the spectrum fragments left are not able to accommodate any new request. An illustration of the operation of the TVWS network is shown in Figure D.1.

The objective is to design how spectrum is assigned at the SM considering the sequential arrival of different bandwidth requests to maximize the overall spectrum utilization of the system.

D.2 PROBLEM FORMULATION AND OPTIMAL SPECTRUM ASSIGNMENT POLICY

The above problem can be considered as a stochastic sequential decision-making problem and can be therefore formulated mathematically under the framework of dynamic programming (see Ref. 45 of Chapter 4). At the time when the tth request arrives, the local SM observes the system state $s_t = \left\{ B_t^{(1)}, \ldots, B_t^{(N)}, b_t \right\}$, which can be characterized by both the available spectrum fragments and the current requested bandwidth. With the observation, the SM has to take an action a_t, where

$a_t \in \{0, 1, \ldots, n, \ldots, N\}$ with $a_t = 0$ representing the current requested b_t is rejected and $a_t = n$ representing b_t is assigned to the nth fragment. When the available spectrum fragment has the capacity to accommodate the incoming request b_t, that is, $\max\{B_t^{(n)}\} > b_t$, the SM has to select one feasible fragment, that is, $a_t \in A(s_t)$, where $A(s_t) = \{n | B_t^{(n)} \geq b_t\}$. Denote $b_{\min}$ as the minimum possible bandwidth of the incoming requests, that is, $b_{\min} = \min\{b | b \in \beta\}$, which can be obtained based on past observations. When the requested bandwidth is greater than any of the available fragments, that is, $b_{\min} \leq \max\{B^{(n)}\} < b_t$, the SM has to reject the current request, that is, $a_t \in A(s_t)$, where $A(s_t) = \{0\}$.

With the action a_t taken, the available spectrum fragments at the time of next spectrum request will be given according to the following system equations:

$$B_{t+1}^{(n)} = \begin{cases} \mathcal{T}, & \max_n\{B_t^{(n)}\} < b_{\min}, \\ B_t^{(n)} - b_t, & B_t^{(n)} \geq b_t \quad \text{and} \quad a_t = n, \\ B_t^{(n)}, & a_t \neq n, \end{cases} \tag{D.1}$$

where $\mathcal{T}$ denotes the termination state when the system cannot accommodate any new request and hence no more action is taken further.

We assume that the tth WSD can achieve a throughput that is proportional to its granted bandwidth b_t if the request is accepted by the SM; otherwise, zero throughput is achieved:

$$r(s_t, a_t) = \begin{cases} b_t, & B_t^{(n)} \geq b_t \text{ and } a_t = n, \\ 0, & \text{otherwise.} \end{cases} \tag{D.2}$$

The objective of our design is to find a policy, which is basically a sequence of functions $\pi = \{\mu_0, \mu_1, \ldots \mu_t, \ldots\}$, $t = 0, 1, \ldots$ mapping each state s_t into control a_t, that is, $a_t = \mu_t(s_t)$ such that overall spectrum utilization of the system is maximized.

$$J^*(s) = \max_\pi \lim_{T \to \infty} E\left[\sum_{t=0}^{T-1} r(s_t, a_t) | s_0 = s\right], \tag{D.3}$$

where $J^*(s)$, $s = \{B_0^{(1)}, \ldots, B_0^{(N)}, b\}$ is the maximum spectrum utilization associated with an initial state s.

According to Ref. 45 of Chapter 4 [3-45], the maximum spectrum utilization $J^*(s)$, $s \in S$, where S is the set of all possible states and is assumed to be finite, is the unique solution satisfying the following Bellman's equation:

$$J^*(s) = \max_{a \in A(s)} r(s, a) + E[J^*(\tilde{s})], \tag{D.4}$$

where $\tilde{s} = \{\tilde{B}^{(1)}, \ldots, \tilde{B}^{(N)}, \tilde{b}\}$ is the next state following the system (D.1) and the expectation is taken with respect to (w.r.t.) the distribution of the request.[1] The

---

[1] If the arrivals of request are independent, the expectation is w.r.t. $Prob\{\tilde{b}\}$; If the arrivals are Markovian, the expectation is taken w.r.t. $Prob\{\tilde{b} | b\}$.

Bellman's equation can be further expanded in the form as shown in (D.5). Note that the maximum throughput is bounded by the total bandwidth of the initial spectrum fragments, that is, $J^*(s_0) \leq \sum_{n=1}^{N} B_0^{(n)}$.

$$J^*\left(B^{(1)}, \ldots, B^{(n)}, \ldots, B^{(N)}, b\right)$$

$$= \begin{cases} 0, & \max\left\{B^{(n)}\right\} < b_{\min}, \\ E\left[J^*\left(B^{(1)}, \ldots, B^{(n)}, \ldots, B^{(N)}, \ldots \tilde{b}\right)\right], & b_{\min} \leq \max\left\{B^{(n)}\right\} < b, \\ \max_{a=n, B^{(n)} \geq b} b + E\left[J^*\left(B^{(1)}, \ldots, B^{(n)} - b, \ldots, B^{(N)}, \tilde{b}\right)\right], & \max\left\{B^{(n)}\right\} \geq b \end{cases}$$

$$(D.5)$$

To calculate the optimal value $J^*(s)$ for $s \in S$, the well-known value iteration method can be applied (see Ref. 45 of Chapter 4) and is described in Algorithm 4.1. The basic idea is to calculate the new value functions $J^{(k+1)}$ based on the value functions computed in the previous iteration $J^{(k)}(s)$ as shown in Ref. 9 of Chapter 4. Then the value functions will converge to the optimal value functions as $k \to \infty$, that is, $J^*(s) = \lim_{k \to \infty} J^{(k)}(s)$. Furthermore, the optimal policy is stationary, that is, $\mu_0 = \ldots = \mu_t = \ldots$, and for any state $s \in S$, it can be computed as in Ref. 10 of Chapter 4 when the algorithm converges.

Appendix E

Calculation for Area-Based WSDB

E.1 METHOD 1: CHANNEL AVAILABILITY FOR AN AREA BASED ON CENTER LOCATION AND RADIUS

In this method, we consider the area as a circle where the SU will provide the center location and the radius of the circle to the WSDB. We describe two schemes to enable the WSDB in checking which channels are available based on these two parameters. When the WSDB approves a channel, the SUs can utilize the channel anywhere within the circle.

The steps for the first scheme are as follows.

E.1.1 M1-Scheme 1-Step 1: Checking the Center Location and Radius

In the first step, after a WSDB receives a request with the center location and radius information from a SU, it will check whether the given center location is within or outside the PU keep-out contour. If the center location is within a PU keep-out contour at a particular channel, then the channel is immediately removed from further consideration and will not be approved for the SU to use. Next, the WSDB will also check the length of the radius provided by the SU as the length of the radius should be limited to prevent the SU from requesting an area that is too big. The radius limit can be a fixed or variable value. For example, we can set the radius limit to be the distance between the center location to the nearest PU location or a fixed distance computed based on SU's maximum transmit power with an appropriate propagation model such as free space model. If the provided radius exceeds this limit, the channel is also immediately removed from further consideration. The two conditions mentioned above are quick ways to remove channels that are unavailable

TV White Space: The First Step Towards Better Utilization of Frequency Spectrum, First Edition.
Ser Wah Oh, Yugang Ma, Ming-Hung Tao, and Edward Peh.
© 2016 by The Institute of Electrical and Electronics Engineers, Inc. Published 2016 by John Wiley & Sons, Inc.

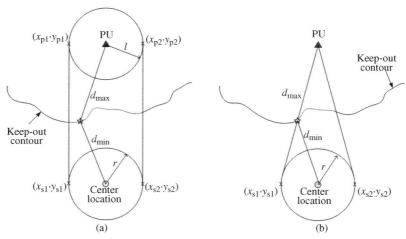

Figure E.1 Enclose an area between PU and center location so as to select a set of PU keep-out contour coordinate points for further processing.

to the SU so as to save computation time. Channels that pass the first step will follow through the remaining steps of the scheme.

E.1.2 M1-Scheme 1-Step 2 (Optional): Finding a Subset of PU Keep-Out Contour Coordinates

The main objective of the WSDB is to make sure that the area requested by the SU does not overlap with the PU keep-out area. To check for all the geographic coordinate points in the PU keep-out contour and make sure that the requesting area does not fall within the PU keep-out contour can be time consuming. To reduce computation time, in step 2, the WSDB selects a subset of PU keep-out contour coordinate points (KOCCPs) for checking instead of checking all of them. The PU KOCCPs that are nearest to the center location should be selected. One way to select the coordinate points is to enclose an arbitrary area between the PU and the center location. The KOCCPs that are inside the enclose area are selected for checking. For example, in Figure E.1a, we use two outer tangents of circles to form an enclose area. The PU KOCCPs that are within the four coordinates, $c_{p1} = (x_{p1}, x_{p1})$, $c_{p2} = (x_{p2}, y_{p2})$, $c_{s1} = (x_{s1}, y_{s1})$, and $c_{s2} = (x_{s2}, y_{s2})$, as shown by the red line in the figure, can be selected. The notation x_i and y_i represent the latitude and longitude of geographic coordinate point c_i. In another example shown in Figure E.1b, an enclosed area between the PU and the center location is formed by the tangent lines from the PU location to the circle area requested by the SU. The PU KOCCPs enclosed by the three points are selected as shown by the red line in the figure and will be used for checking in the next step.

Another method to select a subset of the PU KOCCPs is shown in Figure E.2. A straight line between the PU location and the center location that intersect with

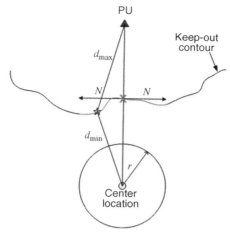

Figure E.2   Select a set of PU keep-out contour coordinate points for further processing based on the intersection of a straight line between PU and center location and the PU keep-out contour.

the PU keep-out contour is drawn. The KOCCP nearest to this intersection point is first chosen, which is shown as a red cross in the figure. From this coordinate point, we can choose N number of coordinate points adjacent to it on both sides to form the subset of PU KOCCPs, represented by the red line in the figure, that need to be checked. If $N = 0$, that means only one coordinate point, the red cross, is selected for checking whether the circle overlaps this point. Step 2 is optional if the computation time is not a concern and the WSDB can choose to check for all the PU KOCCPs instead of a subset of it.

E.1.3 M1-Scheme 1-Step 3: Determine the Area of Circle Is Outside of PU Keep-Out Area

The next step is to decide whether the requesting circle area is outside of the PU keep-out area. We look at two methods in determining whether the requesting circle area is outside of the PU keep-out area.

The first method is to first find the maximum distance among the PU KOCCPs to the PU location. Define $f(c_1, c_2)$ as the function of the distance between two coordinate points, c_1 and c_2. We need to solve

$$d_{max} = \operatorname*{argmax}_{c_i \in C} f(c_i, c_P), \qquad (E.1)$$

where c_p is the coordinate point of the PU location and C is the set of coordinate points from the keep-out contour. If step 2 is not performed, then C contains all the PU KOCCPs. When step 2 is performed and a small number of PU KOCCPs are selected, the above problem can be easily solved by exhaustive search. However, if the number of elements in C is large but the coordinate points are still contained

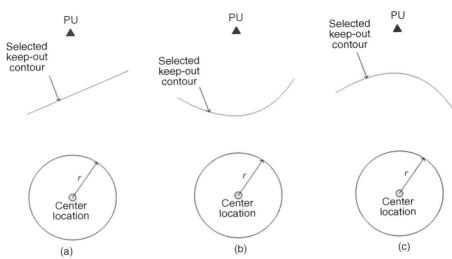

Figure E.3   Three possible shapes of the selected contour line when the selected contour line is a small portion of the whole PU keep-out contour line.

within a limited area as shown in step 2, some efficient algorithms can be implemented by making some reasonable assumptions. For example, the PU keep-out contour line should be relatively smooth since it is derived from propagation models. Hence, if the selected set of coordinate points from the keep-out contour line is a small portion of the whole keep-out contour line, the selected contour line is likely to be a relative straight line, concave line, or convex line as shown in Figure E.3. Efficient methods can be used to solve Equation E.1 based on these properties. For example, if the selected contour line is a straight line or convex line, the coordinate point that has the maximum distance is likely to be at one of the ends of the line. However, even if the selected contour line is concave, which the coordinate point that has the maximum distance is around the middle of the line, efficient search methods such as Golden section search (see Ref. 19 of Chapter 4) can be applied to find that point efficiently. An example program to determine the coordinate point with the maximum distance using Golden section search based on the assumption that the selected contour line is either a straight line, concave line, or convex is provided as below.

Assume that M1-scheme 1-step 2 is performed and the subset of PU KOCCPs that are selected are from c_{pu1} to c_{puk} in sequence where c_{pu1} and c_{puk} are at both ends of the selected contour line and k is the number of coordinate points chosen from the selected contour line. Hence, $C = [c_{pu1}, c_{pu2}, \ldots, c_{puk}]$ in (E.1).

The method to determine the coordinate point with the maximum distance is as follows. First, the distances between the PU location to three points, c_{pu1}, c_{puk}, and $c_{pu(k/2)}$ are determined. If both c_{pu1} and c_{puk} have greater distances than $c_{pu(k/2)}$, which mean the selected line is likely to be convex with two local maximum points at the ends of the line. We split the select contour line into two portions. One portion is from c_{pu1} to $c_{pu(k/2)}$ and the other portion is from $c_{pu(k/2+1)}$ to c_{puk}. Golden section search algorithm is applied on both portions to find the coordinate point that has the maximum distance in each portion. The coordinate point that has the greater distance between the two portions will be determined as the solution.

If $c_{pu(k/2)}$ has a greater distance than c_{pu1} or/and c_{puk}, which means the selected contour line is likely to be concave with one maximum point, the Golden section search algorithm will be applied once between c_{pu1} and c_{puk} to find the coordinate point with the maximum distance. An example of the Golden section search program that finds the coordinate point with the maximum distance between c_{pu1} and c_{puk} is given below where c_p is the PU location.

```
Program: Golden section search to find the coordinate point
with the maximum distance.

Input: C, cₚ
k=length(C);
Tₘᵢₙ=1; Tₘₐₓ=k;
n=(√5-1)/2;
T₁=Tₘᵢₙ+floor((1-n)*( Tₘₐₓ-Tₘᵢₙ));
dist1=f(C(T₁),cₚ);
T₂=Tₘᵢₙ+ceil(n*( Tₘₐₓ-Tₘᵢₙ));
dist2=f(C(T₂),cₚ);

while (Tₘₐₓ-Tₘᵢₙ)>1
if dist1 > dist2
        Tₘₐₓ=T₂;
        T₂=T₁;
        dist2=dist1;
        T₁=Tₘᵢₙ+floor((1-n)*( Tₘₐₓ-Tₘᵢₙ));
        dist1=f(C(T₁),cₚ);
else
        Tₘᵢₙ=T₁;
        T₁=T₂;
        dist1=dist2;
        T₂=Tₘᵢₙ+ceil(n*( Tₘₐₓ-Tₘᵢₙ));
        dist2=f(C(T₂),cₚ);
end
end
if dist1 > dist2
output=dist1;
else
        output=dist2;
end
```

Simulated annealing algorithm (see Ref. 20 of Chapter 4) can also be used to determine the coordinate point that has the maximum distance without the assumptions.

After d_{max} is found, the WSDB will check whether the distance between the PU and the center location is greater than or equal to $d_{max} + r$, where r is the radius of the circle given by the SU. If true, the channel is available for the SUs to operate within the area inside the circle, else the channel is unavailable.

The second method to determine whether the requesting circle is outside of the PU keep-out area is provided. First, we find the minimum distance among the PU KOCCPs to the center location. We need to solve

$$d_{min} = \operatorname*{argmin}_{c_i \in C} f(c_i, c_s), \tag{E.2}$$

where c_s is the coordinate point of the given center location. The method to solve Equation E.2 is similar to Equation E.1. If step 2 is performed and the coordinate points in C is within a limited portion of the PU keep-out contour, then efficient search methods can also be used by assuming that the selected contour line is smooth and has shapes shown in Figure E.3. After d_{min} is found, the WSDB will check whether it is greater than or equal to r. If true, the channel is available for the SUs to operate within the area inside the circle, else the channel is unavailable. The process of Scheme 1 for method 1 that makes decisions based on center location and radius is summarized in Figure E.4.

Next we present the steps for the second scheme which are as follows.

E.1.4 M1-Scheme 2-Step 1: Checking the Center Location and Radius

The first step of Scheme 2 is the same as Scheme 1. Please refer to Section E.1.1.

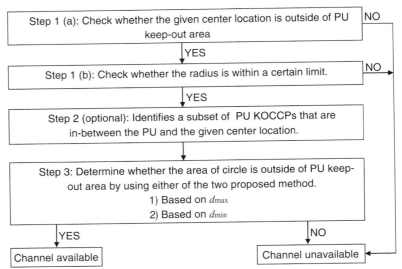

Figure E.4 The flowchart of Scheme 1 to determine whether a channel is available for an area defined by a center location and radius.

E.1.5 M1-Scheme 2-Step 2: Determine a Set of Location Check Points

The next step is for the WSDB to identify a set of location points denoted as location check points (LCPs). These LCPs will be determined using the center point and the radius. As shown in Figure E.5, the LCPs are chosen from the circumference of the circle. The LCPs can be chosen from the whole circumference or a section of the circumference. For example, a LCP can be chosen at every 1° in direction from the whole circumference forming a set of 360 LCPs. It is possible to reduce the number of LCPs by choosing them from a section of the circumference that is in a limited degree range from the direction of the PU. For example, in Figure E.5, the LCPs can be chosen within $\theta_1°$ from the direction of the PU, which can reduce the computation time in the next step.

E.1.6 M1-Scheme 2-Step 3: Determine If All the LCPs Are Outside of the PU Keep-Out Contour

The last step of this scheme is to determine if all the LCPs chosen in step 2 are outside of the PU keep-out contour. When all the LCPs are outside of the PU

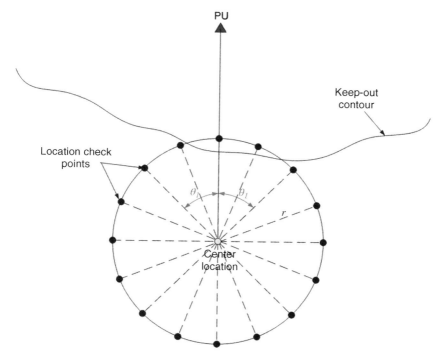

Figure E.5   Identify a set of location check points based on the center location and radius. When all the LCPs in the set are outside of the PU keep-out contour then the channel will be available.

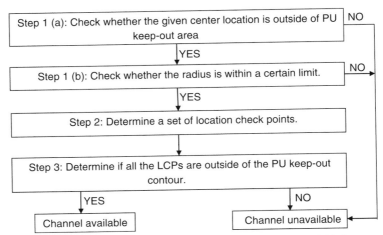

Figure E.6   The flowchart of Scheme 2 to determine whether a channel is available for an area defined by a center location and radius.

keep-out contour, the channel will be determined by the WSDB as available to the SUs to operate within the area inside the circle, else the channel is unavailable. Conventional WSDB (see Refs 13–15 of Chapter 4) is able to determine whether a single location is outside of the PU keep-out contour, hence we can use conventional WSDB to determine whether all the LCPs are outside of the PU keep-out contour. The process of Scheme 2 for method 1 that makes decisions based on center location and radius is summarized in Figure E.6.

E.2   METHOD 2: CHANNEL AVAILABILITY FOR AN ARBITRARY AREA BOUNDED BY A SERIES OF LOCATION INFORMATION

In this method, the SU will provide a sequence of location information that serves as a boundary that the SUs will operate within. One simple way to determine whether a channel is available is to check whether the provided coordinate points (PCPs) that formed the boundary of the requesting area are all outside of the PU keep-out contour. If they are all outside, the channel will be declared available for the SU to use within the bounded area. This method is simple, however, it has shortcomings that will be pointed out subsequently. In Figure E.7a, the red crosses are the PCPs and it can be seen that all of them are outside of PU keep-out contour. Hence, if based on these coordinate points, the channels are available to the SU that makes this method failed in protecting the PU network from the SUs. This scenario fails because the requesting area is larger than the PU keep-out area but even if we constraint the requesting area to be smaller than the PU keep-out area as shown in Figure E.7b, it is still possible for the requesting area to overlap with the PU

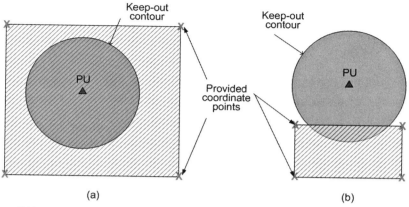

Figure E.7   Possible scenarios of failures if decisions on channel availability are purely based on whether boundary locations are outside of PU keep-out contour.

keep-out area based on the PCPs, if the PCPs are far apart. To overcome these shortcomings, we describe two schemes for this method.

E.2.1   M2-Scheme 1

The first scheme adds an additional check to determine whether there are any PU KOCCPs within the area bounded by the PCPs after checking that the PCPs are outside of the PU keep-out contour. If none of the PU KOCCP is inside the bounded area, the WSDB will decide that the channel is available to the SUs to operate within the bounded area. This process is not complicated and programs to determine whether there is any PU KOCCPs inside an area can be done by just a few lines of codes, especially if the bounded area is of a regular shape. For example, we show the codes for this process when the bounded area is a rectangular shape as below.

Assume that a SU provides four coordinate points that form a rectangle with $c_{su1} = (x_a, y_a)$, $c_{su2} = (x_a, y_b)$, $c_{su3} = (x_b, y_b)$, and $c_{su4} = (x_b, y_a)$, where $x_b > x_a$ and $y_b > y_a$, to the WSDB. The WSDB would have known the coordinate points of the PU KOCCPs through precomputation based on PU transmit parameters and propagation models. Denote the PU KOCCPs as $c_{pu1} = (x_{pu1}, y_{pu1})$, $c_{pu2} = (x_{pu2}, y_{pu2})$, . . . , $c_{puM} = (x_{puM}, y_{puM})$, where M is the total number of PU KOCCPs that formed the boundary of the PU keep-out contour. Let $X_{PU} = [x_{pu1}, x_{pu2}, . . . , x_{puM}]$ and $Y_{PU} = [y_{pu1}, y_{pu2}, . . . , y_{puM}]$, an example program to check if any of the PU KOCCPs falls inside the SU PCPs is given below. If the program output $i = 0$, means there is no PU KOCCP inside the rectangular area. If the program output $i = 1$, then at least one PU KOCCP is inside the rectangular area. So if $i = 0$, the WSDB will approve the channel to be used within the area.

```
Program: Determine any PU keep-out contour coordinate points
inside a rectangular area.

Input: xₐ, x_b, yₐ, y_b, X_PU, Y_PU
M=length(X_PU);
i=0; j=1;
while(i=0)&(j<=M)
        if (X_PU(j) >= xₐ) & (X_PU(j) <= x_b)
    if (Y_PU(j) >= yₐ) & (Y_PU(j) <= y_b)
    i=1;
    end
        end
j=j+1;
end

Output: i
```

It is also possible to include a step similar to *M1-Scheme 1-Step 2* to select a subset of PU KOCCPs for checking whether any of these points fall inside the bounded area to reduce the computation time. By including the second check, if there is no PU KOCCPs within the bounded area, there will be no overlap between the SU requesting area and the PU keep-out area that the first check alone cannot guarantee. The process of this scheme is summarized in the flowchart shown in Figure E.8.

E.2.2 M2-Scheme 2

Since checking the PCPs may fail if the requesting area is larger than the PU keep-out area and/or the PCPs are too far apart, we can make some constraints to limit them. First, add a constraint to limit the size of the requesting area using the distances among the PCPs. For example, a step can be added such that when the maximum distance among the PCPs exceeds a threshold value, the channel will not be available to the SU. Second, add a constraint to limit a PCP from being too far from

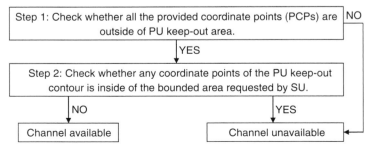

Figure E.8   The flowchart of Scheme 1 to determine whether a channel is available for an area defined by a series of coordinate points provided by a SU.

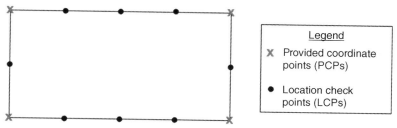

Figure E.9 Location check points are determined along the connecting lines of the provided coordinate points for checking.

its connecting PCPs. For example, a step can be added such that the distance between every PCP and its connecting PCP must be smaller than a threshold value. When the distance between a PCP and its connecting PCP exceeds the threshold value, the WSDB has two options. One option is to straight away determine that the channel is not available for the SU. The other option is that the WSDB will identify LCPs between the PCPs as intermittent connecting points so that the distance between every PCP/LCP and its connecting PCP/LCP will be within the threshold value. An example is shown in Figure E.9, where LCPs are determined along the connecting lines of the PCPs that form the boundary of the requesting area. When all the PCPs and LCPs are determined to be outside of the PU keep-out contour, the channel will be determined as available for the SU to operate within the area. The process of this scheme is summarized in the flowchart shown in Figure E.10.

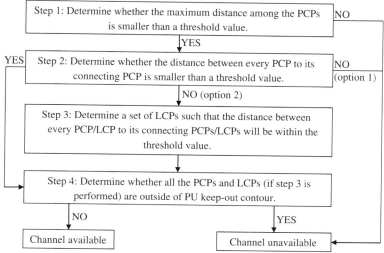

Figure E.10 The flowchart of Scheme 1 to determine whether a channel is available for an area defined by a series of coordinate points provided by a SU.

Appendix F

Embedded Broadcast WSDB

F.1 TELETEXT-BASED BROADCAST WSDB

Teletext is an existing function in analogue TV broadcasting. There is no additional hardware cost in TV broadcast stations if we use the existing infrastructure to broadcast WSDB. The block diagram of the system using teletext is shown in Figure F.1. The WSDB data are first encrypted and then embedded in the vertical blinking period of the TV signal similar to the teletext signal. The TV signal including WSDB data is sent to TV transmitter for broadcast. The same process is done in every TV tower connected to the WSDB.

F.2 DVB- OR HBB-BASED BROADCAST WSDB

There are a few ways to embed WSDB data in a DVB transmission. First, the WSDB data can be embedded as PSI/SI. This is the straightforward method. For DVB-T2 system, the data can also be embedded in the other use period in the future

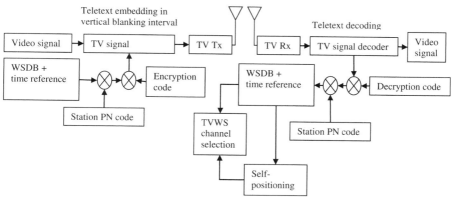

Figure F.1 Block diagram of signal embedding and decoding for teletext-based broadcast WSDB.

TV White Space: The First Step Towards Better Utilization of Frequency Spectrum, First Edition.
Ser Wah Oh, Yugang Ma, Ming-Hung Tao, and Edward Peh.

313

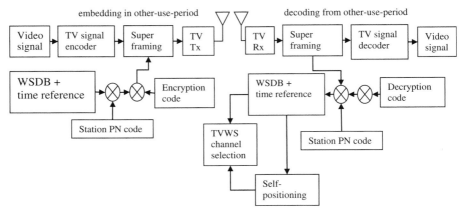

Figure F.2   Block diagram of signal embedding and decoding for DVB-T2-based broadcast WSDB.

extension frame (FEF) in its superframe. In a hybrid broadcast broadband (HBB) TV system, there is strong text transmission function besides the normal TV signals. This text transmission function can be used to broadcast WSDB information.

In this section, we discuss the single-frequency network (SFN) DVB-T2 system. In a SFN, a WSD can receive the signal of a TV channel from multiple TV towers. Figure F.2 shows the block diagram of the signal embedding and decoding in DVB-T2 system; Figure F.3 shows the system scenario of broadcast WSDB in SFN DVB-T2. The WSDB and time reference signals are embedded in the other

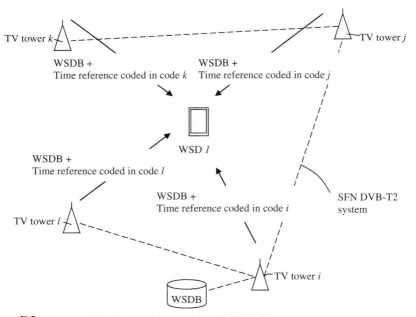

Figure F.3   Broadcast WSDB TVWS system with SFN DVB-T2.

use period of FEF of the superframe. The WSDB data and time reference data are coded by TV tower PN code and encryption code, respectively. TV tower location, WSDB, and time reference data are transmitted simultaneously from multiple TV towers along with the TV signals. The dedicated pseudo-noise (PN) code for each TV tower is assigned and known to the WSD. The WSD can decode the WSDB and time reference signal from different TV towers by using the correct PN codes.

The self-position of the WSD is estimated using the time reference signals from different towers. This self-positioning is suitable for both indoor and outdoor environments.

In the case that the WSD does not carry out self-positioning through the time reference signal, the WSD can also equip an additional GPS device for positioning in an outdoor environment.

Having received the WSDB and carried out self-positioning, the WSD can find the available TV channels at its own position and select one for TVWS communications.

In the embedding approach based on FEF, the WSD does not need to recover the TV signal but only decodes the other use period in FEF.

F.3 DAB- OR HD RADIO-BASED BROADCAST WSDB

The WSDB and time reference signals can also be transmitted through audio broadcasting DAB or HD. In DAB, it can be embedded as FIC/PAD. In HD radio, it can be embedded at L4 encoding. The signal embedding and decoding methods in the HD radio case are described in Figure F.4. The broadcast WSDB system with HD radio is shown in Figure F.5. The WSD has to have the receivers of each radio frequencies and/or modulations. The WSDB is received from one of the stations. Similarly, the self-position can be estimated by using the time reference signals received from multiple stations (at least three) from different locations.

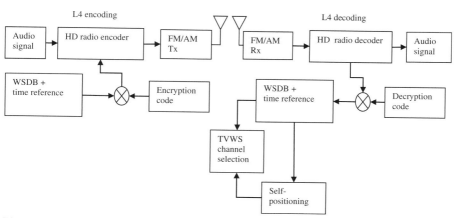

Figure F.4   Block diagram of signal embedding and de-embedding for HD radio-based broadcast WSDB.

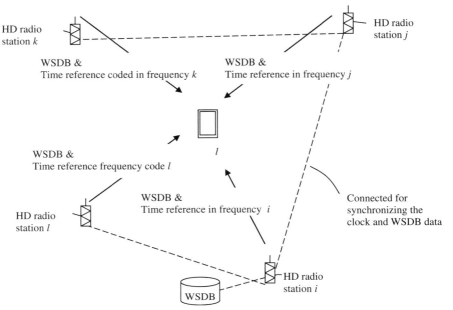

Figure F.5 HD radio-based broadcasting TVWS system.

Appendix G

Revenue Maximization of WSDB-Q[*]

G.1 MAXIMIZING REVENUE OF WSDB PROVIDER

WSDB is able to know the total number of WSDs as well as the number of WSDs accessing HPCs since WSDs need to inquire WSDB prior to communication. The mathematical formulation of this problem is given as

$$\max_{f} F = n_H f,$$

$$\text{s.t.} \quad n_H + n_N = N,$$

$$0 \le f \le f_M, \tag{G.1}$$

where N is the total number of WSDs. First, we determine the number of WSDs that will use HPCs for a given fee $\check{f}$ between 0 and f_M. Based on the reward criterion that a WSD will join HPCs if

$$\frac{f_M}{R_T - 1}(R - 1) \ge \check{f}, \tag{G.2}$$

we obtain

$$n_H \le \frac{m_H N + m_H m_N \alpha\left(\check{f}\right)}{m_H + m_N}, \tag{G.3}$$

where

$$\alpha\left(\check{f}\right) = \frac{\ln\left(\check{f}(R_T - 1) + f_M\right) - \ln(f_M)}{\ln(1 - \tau)}. \tag{G.4}$$

[*] *Source*: Pricing model for TV white spaces channels with different priorities. *IEEE Wireless Communications Letters*, Vol. 4, No. 4, August 2015.

TV White Space: The First Step Towards Better Utilization of Frequency Spectrum, First Edition.
Ser Wah Oh, Yugang Ma, Ming-Hung Tao, and Edward Peh.

Based on the QoS criterion that a WSD will join HPCs if $P_H \geq \overline{P}_H$, we will obtain

$$n_H \leq m_H \left(\frac{\ln\left(\overline{P}_H\right)}{\ln\left(1 - \tau\right)} + 1 \right) = \overline{n}_H. \tag{G.5}$$

Therefore, from Equations G.4 and G.5, the number of WSDs that will use HPCs at a given fee $\overset{\vee}{f}$ is given as

$$n_H\left(\overset{\vee}{f}\right) = \max\left\{ 0, \min\left\{ \frac{m_H N + m_H m_N \alpha\left(\overset{\vee}{f}\right)}{m_H + m_N}, \overline{n}_H \right\} \right\}, \tag{G.6}$$

where x means rounding down x to the nearest integer. The total fee collected by WSDB provider when it sets its fee to $\overset{\vee}{f}$ is given as

$$F\left(\overset{\vee}{f}\right) = n_H\left(\overset{\vee}{f}\right)\overset{\vee}{f}. \tag{G.7}$$

From Equation G.6, $n_H(f) = \overline{n}_H$ when

$$f \leq \frac{f_M(1 - \tau)^{\frac{m_H + m_N}{m_H m_N}\overline{n}_H - N/m_N} - f_M}{R_T - 1}. \tag{G.8}$$

This means that if a WSDB provider sets a fee lower than the right-hand term in Equation G.8, it will not attract further more WSDs to use HPCs as the number of WSDs in HPCs is already saturated and could not provide the required QoS if more WSDs are to join in. Hence, the minimum fee that a WSDB provider should set based on the QoS constraint is given as

$$f_{QoS} = \max\left\{ 0, \frac{f_M(1 - \tau)^{\frac{m_H + m_N}{m_H m_N}\overline{n}_H - N/m_N} - f_M}{R_T - 1} \right\} \tag{G.9}$$

unless $f_{QoS} > f_M$, which is when

$$N > \frac{m_H + m_N}{m_H}\overline{n}_H - \frac{m_N \ln\left(R_T\right)}{\ln\left(1 - \tau\right)}. \tag{G.10}$$

In this case, it is straight forward that the optimal fee f^* is f_M and the optimal revenue collected by the WSDB provider is $F^* = \overline{n}_H f_M$. This is because reducing the fee to be less than f_M will not increase the number of WSDs in HPCs.

For $N \geq \frac{m_H + m_N}{m_H}\overline{n}_H - \frac{m_N \ln\left(R_T\right)}{\ln\left(1 - \tau\right)}$, the optimization problem becomes

$$\max_f F(f) = f\,\frac{m_H N + m_H m_N \alpha\left(\overset{\vee}{f}\right)}{m_H + m_N},$$

$$\text{s.t.} \quad f_{QoS} \leq f \leq f_M. \tag{G.11}$$

To find the optimal f^* that will maximize Equation G.11, we define a new function to approximate the total revenue collected by WSDB provider in Equation G.11 by

$$\tilde{F}(f) = f\left(\frac{m_{\mathrm{H}}N + m_{\mathrm{H}}m_{\mathrm{N}}\alpha\left(\overset{\vee}{f}\right)}{m_{\mathrm{H}} + m_{\mathrm{N}}}\right). \tag{G.12}$$

The equation $\tilde{F}(f)$ is a concave function as its second derivative is always less than zero for all f, given as

$$\frac{\partial^2}{\partial f^2}\tilde{F}(f) = \frac{m_{\mathrm{H}}m_{\mathrm{N}}(R_{\mathrm{T}} - 1)}{(m_{\mathrm{H}+}m_{\mathrm{N}})\ln(1-\tau)}\left(\frac{1}{f(R_{\mathrm{T}})+f_{\mathrm{M}}} + \frac{f_{\mathrm{M}}}{(f(R_{\mathrm{T}}-1)+f_{\mathrm{M}})^2}\right) < 0 \forall f. \tag{G.13}$$

The optimal $\tilde{f}^*$ with Equation G.12 as the objective function occurs when $\frac{\partial^2}{\partial f^2}\tilde{F}(f) = 0$, which can be easily solved by basic descent algorithms. The derivative of $\tilde{F}(f)$ is given as

$$\frac{\partial^2}{\partial f^2}\tilde{F}(f) = \frac{m_{\mathrm{H}}m_{\mathrm{N}}}{m_{\mathrm{H}+}m_{\mathrm{N}}}\left(\frac{N}{m_{\mathrm{N}}} + \alpha(f) + \frac{f(R_{\mathrm{T}} - 1)}{(f(R_{\mathrm{T}} - 1)+f_{\mathrm{M}})\ln(1-\tau)}\right). \tag{G.14}$$

However, the value of $\tilde{f}^*$ may not be the optimal solution for Equation G.11, as the optimal solution of Equation G.11 must either be f_{M} or a f value that makes $\left(m_{\mathrm{H}}N + m_{\mathrm{H}}m_{\mathrm{N}}\alpha\left(\overset{\vee}{f}\right)\right)/(m_{\mathrm{H}} + m_{\mathrm{N}})$ an integer. The proof is as follows.

Let f^i be a value such that $(m_{\mathrm{H}}N + m_{\mathrm{H}}m_{\mathrm{N}}\alpha(f^i))/(m_{\mathrm{H}} + m_{\mathrm{N}})$ is an integer and f^{i+k} be a value such that

$$\frac{m_{\mathrm{H}}N + m_{\mathrm{H}}m_{\mathrm{N}}\alpha(f^i)}{m_{\mathrm{H}} + m_{\mathrm{N}}} + k = \frac{m_{\mathrm{H}}N + m_{\mathrm{H}}m_{\mathrm{N}}\alpha(f^{i+k})}{m_{\mathrm{H}} + m_{\mathrm{N}}}. \tag{G.15}$$

For a given $\hat{f}$ that is $f^i < \hat{f} < f^{i+1}$, we have $F(\hat{f}) < F(f^{i+1})$ since

$$\frac{m_{\mathrm{H}}N + m_{\mathrm{H}}m_{\mathrm{N}}\alpha(\hat{f})}{m_{\mathrm{H}} + m_{\mathrm{N}}} = \frac{m_{\mathrm{H}}N + m_{\mathrm{H}}m_{\mathrm{N}}\alpha(f^{i+1})}{m_{\mathrm{H}} + m_{\mathrm{N}}}. \tag{G.16}$$

Hence, any f value between f^i and f^{i+1} will not be optimal. Next, we let $f^{i+k_{\max}}$ be the largest value of f that makes $(m_{\mathrm{H}}N + m_{\mathrm{H}}m_{\mathrm{N}}\alpha(f))/(m_{\mathrm{H}} + m_{\mathrm{N}})$ an integer. For a given $\hat{f}$ that is $f^{i+k_{\max}} < \hat{f} < f_{\mathrm{M}}$, we have $F(\hat{f}) < F(f_{\mathrm{M}})$ since

$$\frac{m_{\mathrm{H}}N + m_{\mathrm{H}}m_{\mathrm{N}}\alpha(\hat{f})}{m_{\mathrm{H}} + m_{\mathrm{N}}} = \frac{m_{\mathrm{H}}N + m_{\mathrm{H}}m_{\mathrm{N}}\alpha(f_{\mathrm{M}})}{m_{\mathrm{H}} + m_{\mathrm{N}}}. \tag{G.17}$$

Therefore, the optimal f^* for Equation G.11 must either be f_M or a f value that makes $(m_{\mathrm{H}}N + m_{\mathrm{H}}m_{\mathrm{N}}\alpha(\hat{f}))/(m_{\mathrm{H}} + m_{\mathrm{N}})$ an integer.

We define

$$\tilde{f}^L = \frac{f_M\left((1-\tau)^{\frac{m_H+m_N}{m_H m_N} \frac{m_H N+m_H m_N \alpha(\tilde{f}^*)}{m_H+m_N} - \frac{N}{m_N}} - 1\right) - f_M}{R_T - 1}, \tag{G.18}$$

and

$$\tilde{f}^U = \min\left\{\frac{f_M\left((1-\tau)^{\frac{m_H+m_N}{m_H m_N} \frac{m_H N+m_H m_N \alpha(\tilde{f}^*)}{m_H+m_N} - \frac{N}{m_N}} - 1\right)}{R_T - 1}, f_M\right\}, \tag{G.19}$$

where x means rounding-off x to the nearest integer. The optimal solution of Equation G.11 is therefore given as

$$f^* = \begin{cases} \tilde{f}^L & F(\tilde{f}^L) > F(\tilde{f}^U) \\ \tilde{f}^U & \text{otherwise} \end{cases}. \tag{G.20}$$

This is because from Equations G.11 and G.12, $\tilde{F}(f) = F(f)$ when $(m_H N + m_H m_N \alpha(f))/(m_H + m_N)$ is an integer, which is also the point where the f values satisfy the optimal criterion of Equation G.11. Among all the f values between f_{QoS} and $\tilde{f}^*$ that satisfy the optimal criterion, $F(\tilde{f}^L)$ will obtain the largest revenue since we have proven $\tilde{F}(f)$ is a concave function. Similarly, among all the f values between $\tilde{f}^*$ and f_{max} that satisfy the optimal criterion, $F(\tilde{f}^U)$ will obtain the largest revenue as $\tilde{F}(f)$ is a concave function. Hence, the optimal solution of Equation G.11 is either $\tilde{f}^L$ or $\tilde{f}^U$, whichever that gives the higher revenue. Finally, the optimal solution of Equation G.11 is f_M when $N > \frac{m_H+m_N}{m_H}\bar{n}_H - \frac{m_N \ln(R_T)}{\ln(1-\tau)}$ and f^* in Equation G.11 when $N \le \frac{m_H+m_N}{m_H}\bar{n}_H - \frac{m_N \ln(R_T)}{\ln(1-\tau)}$.

Index

TV White Space: The First Step Towards Better Utilization of Frequency Spectrum, First Edition.
Ser Wah Oh, Yugang Ma, Ming-Hung Tao, and Edward Peh.
© 2016 by The Institute of Electrical and Electronics Engineers, Inc. Published 2016 by John Wiley & Sons, Inc.

THE COMSOC GUIDES TO COMMUNICATIONS TECHNOLOGIES

Nim K. Cheung, *Series Editor*
Thomas Banwell, *Associate Editor*
Richard Lau, *Associate Editor*